STATISTICAL MODELING USING 1-2-3

RELEASE 2

William T. Cloake

MANAGEMENT INFORMATION SOURCE, INC.

COPYRIGHT

Copyright © 1988 by Management Information Source, Inc.
1107 N.W. 14th Avenue
Portland, Oregon 97209
(503) 222-2399

First Printing

ISBN 0-943518-32-6

Library of Congress Catalog Card Number: 87-12347

dBASE is a trademark of Ashton-Tate
Lotus 1-2-3 and PrintGraph are trademarks of Lotus Development Corporation
PC-DOS is a trademark of IBM Corporation
WordStar is a trademark of MicroPro International Corporation

DEDICATION

To my sons William (a.k.a. Joe Willie), who plays quarterback and is a Student Senator at Fullerton College, and Sean, a future engineer who plays bass clarinet and runs track for Ontario High School. I hope this book gives them as much pride as they give me.

ACKNOWLEDGEMENTS

A project like this is never completed without the help of many people. There are a few I must thank individually.

Without the support of my wife Annette, this book would never have been completed. She has proofread, picked up, delivered, and helped in countless ways. She can easily drive to Federal Express blindfolded.

My son Sean took my picture for the cover and has spent literally hundreds of hours keeping me company. Every author should have a morale officer like Sean on staff.

My son William not only proofread most of this book twice, but took on the laborious task of cross-checking all of the macro printouts. Without his help, readers would find hundreds of errors.

I must thank my "telephone friend," Chris Williams, who was instrumental in starting this project.

Last, and certainly not least, are Kim Thomas and her staff at MIS:Press. Without Kim's help and creative input, this book would have none of the polish and style you see here. Kim not only dotted i's and crossed t's (the smallest part of an editor's job), but oversaw the cover design, the book's layout and typesetting, and the general appearance of the finished book. In her spare time, she answered questions about every aspect of the publishing process. Now I know for sure what a technical editor is. Thanks, Kim, for another job well done.

TABLE OF CONTENTS

Preface xi
Introduction xvii

Part I: Warm-Up Exercises 1
   Why a Standard Output? 3

Chapter 1: @ Functions and Ranges 5
   The @ Functions 6
   Range Descriptions 8
   Example Problem 14
      Values 14
      Solution 14
   Summary 16
   Exercises 16
      Case Study One: Widget Sales 16
      Answers 17
   List of Functions 23

Chapter 2: /Data Base Commands 25
   Data Base Commands 27
      Fill 28
      Data Sort 32
   Data Query Operations 36
      /Data Query Find 37
      /Data Query eXtract 41
      Additional Data Commands 44
   Summary 44
   Exercises 45
      Case Study Two: Multiple Office Sales Department 45
      Answers 48

Chapter 3: Macros 55
   When to Use a Macro 56
   Creating and Saving a Macro 57
   Creating a Simple Macro 59
      Second Macro Example 61
   Debugging a Macro 64
   Input to a Macro 64
   The /X Commands 64

Summary	67
Exercises	67
Answers	68

Chapter 4: File Operations 71

Getting Data to a Disk	72
File Extractions	74
File Printing	77
Retrieving Data from a Disk	78
File Combine	78
Importing Data	79
Summary	83
Exercises	83
Case Study Two (Continued)	83
Answers	84

Chapter 5: Creating and Saving Graphs 89

Graph Types	92
Line Graphs	93
Bar Graphs	95
Stacked Bar Graphs	96
Pie Graphs	97
X-Y Graphs	98
Reset	100
Save	100
Options	101
Legend	101
Format	101
Titles	102
Grid	103
Scale	104
Naming a Graph	106
Summary	108
Exercises	108
Case Study Two: Conclusion	108
Answers	109

Part II: Single-Variable Analysis 119

Chapter 6: File Management 123

Method	125

        Application 126
        Output 126
        Required Commands 127
        Template Organization 127
        Menu 129
        Macro Functions 130
             Variable Labels 130
             Screen Labels and Marquis Listing 132
             The Autostart Macro 133
             Macro Menu 134
        Submacro Operation 135
             Extract 135
             Count 136
             Restart 137
             Erase 137
             Combine 138
             Import 139
        Cell Range Names 140
        Putting Them Together 140
        Summary 143
        Exercises 143
             Answers 144
        Template Listing 146

Chapter 7: Small Sample Statistics 151
        Methods and Applications 154
             Min, Max, Average 154
             Median 154
             Mode 157
        Output 160
        Required Commands 160
        Menus 161
             Main Menu 163
        Macro Descriptions 164
             Load 164
             Median/Mode 166
             Print 171
             Graph 172
        Range Names 173
        Operation 173
        Summary 175
        Exercises 176

  Case Study Three: Heights of Students 176
   Answers 177
 Template Listing 179

Chapter 8: Descriptive Statistics 189
 Menus 193
  Autostart 193
 Layout of the Template 195
 Methods and Applications 196
  Mean and Standard Deviation 197
  Unbiased Estimator 199
  Moments 201
  Skewness 201
  Kurtosis 201
  Histograms 202
 Outputs 203
 Commands Used 203
 Macro Descriptions 204
  Autostart 204
  Menu Functions 204
  Load 207
  Unload 210
  Utility Subroutines 210
  Calculations for Frequency Distribution 211
 Range Names 212
 Summary 213
 Exercises 214
  Answers 214
 Template Listing 216

Chapter 9: Graphic Methods 227
 Menus 231
  Print 232
  Graph 233
  Cursor 233
  Adjust 233
 Methods and Applications 234
 Outputs 237
 Commands Used 242
 Macro Functions 243
  Line Graph of Data Values 234
  Graph Save Subroutine (Part I) 244

  Graph Save Subroutine (Part II) 244
  Graph Save Subroutine (Part III: Operator Instructions Menu) .. 245
  Calculations for Frequency Distribution 245
  Binlim Subroutine 246
 Range Names 246
 Summary 247
 Exercises 247
  Answers 248
 Template Listing 251

Part III: Two-Way Analysis 265

Chapter 10: Two-Way Analysis 269
 Menus 271
  Test 273
  File 273
  Cursor and Adjust 273
 Spreadsheet Layout 274
 Methods and Applications 275
  Z Test 277
  T Test 279
  F Test 279
  Regression 280
  Project 281
 Output 281
  Sample Statistics 281
  Regression Information 282
  Hypothesis Test Result 283
 Commands Used 283
 Macro Descriptions 284
  Load Macro 284
  Load Y Macro 286
  Unload Macro 286
  Table Subroutine 287
  Regression Subroutine 287
  Z Test Subroutine 289
  T Test Subroutine 290
  F Test Subroutine 290
 Range Names 290
 Examples 291
  Manufacturing Examples 291

Summary 294
　　　Exercises 295
　　　　　Case Study Four: Manufacturing 295
　　　　　Answers 296
　　　Template Listing 298

Chapter 11: Two-Way Graphic Methods 309
　　　Menus 311
　　　Methods and Applications 313
　　　　　X-Y and Y-X Scatter Plots and Regression 314
　　　　　Smoothing 316
　　　　　Gaussian 317
　　　Output 322
　　　Commands Used 322
　　　Macro Descriptions 323
　　　　　Unload Macro 323
　　　　　Print Summary Subroutine 323
　　　　　Test Results Subroutine 324
　　　　　Scratchpad Subroutine 324
　　　　　Smooth Subroutine 325
　　　　　Xscat and Yscat Subroutines 325
　　　　　Gauss Subroutine 326
　　　　　Regression Graph Subroutine 327
　　　Range Names 327
　　　Summary 328
　　　Exercises 328
　　　　　Answers 329
　　　Template Listing 332

Chapter 12: Nonparametric Methods 349
　　　Menus 352
　　　　　Main Menu 353
　　　　　File Menu 353
　　　　　Test Menu 353
　　　　　Print Menu 353
　　　Methods and Applications 353
　　　　　Runs Test 353
　　　　　Sign Test 356
　　　　　Case Study Five: Manufacturing 356
　　　　　Mann-Whitney Test 358
　　　　　Case Study Six: Softball 358
　　　Commands Used 362

Macro Descriptions 363
    Load Subroutine 363
    Print Subroutine 363
    Const Subroutine 364
    Unload Subroutine 365
    Runs Test 366
    Sign Test 366
    Mann-Whitney Test 367
    Rank Sum Subroutine 368
Range Names 370
Summary 370
Exercises 371
    Case Study Seven: Education 371
        Answers 372
Template Listing 375

Part IV: User's Guide 383

Chapter 13: User's Guide 385
The Composite Menu 386
Installing the Program 388
    Tips on Customizing 389
    Disk File Options 391
Running the Template Package 394
    Data File Creation 394
    Small Sample Techniques 396
    Large Sample Techniques 397
    Two-way Analysis 399
    Nonparametric Methods 400
Examples 400
    Case Study Eight: Vendor Qualification 400
    Case Study Nine: Marketing 406
    Case Study Ten: Engineering 409

Appendices 415
Appendix A: Tables 417
Appendix B: Enhancing Output with Printer Options 423
    Print Options 425
    Printer Commands 426
    Setup Strings 431
    Macro Control of Printer Options 432

    Labels for Print Example 433
Appendix C: Writing Software for Importing Data into Lotus 435
    Basic Program Description 436
    Macro Description 438
    Installing the Macro 440
    Using the BASIC Program 441
    BASIC Program Listing 442
    Main Program Body 443
    File Translation Macro 449
Appendix D: Macro Menus and Marquis 451
    The Game of NIM 451
    The Screen Layout 452
    NIM Screen Layout 453
    Macro Program Body 453
    The Subroutines 457
    Running the Macro 460
    The Statistics Package Menu 461
    The Marquis 462
    Menus 463
    Range Names 466
Appendix E: Customizing Printgraph Outputs 467
    Printgraph Functions 467
    Sample Graph 470
    Manual Print Options with Printgraph 472
Appendix F: Creating Macro Listings 477
    File Printing 477
    Importing and Sorting .PRN Files 480
Appendix G: DOS Utilities You Need to Know 487
    File Manipulation 487
        Format 487
        CHKDSK 489
        DISKCOPY 489
        COPY 490
        DELETE and ERASE 492
    Advanced Directory Utilities 493
        DIR 493
        Directory Trees 494
        TREE 495
    Subdirectory Operations 495
        MKDIR 495
        CHDIR 496
        RMDIR 497

  Path 498
 Additional Utility Functions 499
  Prompt 499
  Attrib 500
  Subst 500

Index...**503**

PREFACE

The route to a Computer Science degree is filled with Math, Engineering, and Quantitative Methods courses, including up to 18 units of Statistics. Those of us who went through such a program were used to Math classes that directly applied to Engineering and Physics courses, laboratory work that paralleled Engineering courses, and Quantitative Methods classes that were unrelated but very practical and clear. Statistics just did not fit the mold. In fact, the net result for me was a simple hypothesis: I'll never use this stuff.

A few years and a few grey hairs later, I found myself running a Quality Assurance department. At once, there were literally thousands of pieces of information on the one hand and business and technical decisions to be made on the other. How could good units be separated from defectives? Which of the parameters we measured could tell us most about our quality? How many inspectors did we need? If output quintupled, would we need five times as many testers?

Clearly, these decisions were complex and did not lend themselves to trivial solutions. Simply rejecting 20% of the units completed, without solving quality problems, did nothing to improve product quality. Neither did it allow better understanding of the accept/reject criteria. Astute customers wanted to know these criteria, along with forecasts of when the product would improve. We found that we had, right in our hands, 30 pieces of information on each unit we built, but we threw that information away because we didn't know what to do with it. It was clear that we rejected units that were good and shipped some that were defective. (Type I and II errors, for those who have been to class.)

As we stepped up production — buying expensive test equipment and hiring ad infinitum — expensive solutions resulted that made no sense from a business standpoint. It became obvious that if yields could be improved in the process, output could increase with the existing equipment and manpower. In short, we began implementing a Statistical Process Control system. My hypothesis about statistics didn't prove true: I *was* using this stuff.

Inspiration

Most of the tests I had acquired over the years were aimed at the theory of hypothesis testing rather than the day-to-day chores of using statistics. I learned more from my boss and other professionals in the QA field than from the courses completed earlier. One fact became clear: there were many statistics texts available but none that offered the professional a way to become productive quickly and build a solid understanding of basic elementary statistics. No assistance was available to expand the capabilities of Lotus 1-2-3 into a real statistics system.

This text fills that void. It can be used in conjunction with any entry-level course in Math, Science, Engineering, or Business or as a stand-alone workbook for students or professionals who need to learn how to reduce data for effective decision making.

Evolution

Early in the development of our Statistical Process Control system, it became clear that computerized data reduction was more necessary than nice. To that end, we developed a simple software set in Fortran on a DEC system. At this point, I began developing my own software package, in BASIC, on my home computer.

Initially, this BASIC language "statpack" emulated the DEC histogram package and gradually was increased to include most of the functions you will find in this book. The statpack is a collection of subroutines and small programs tied together by application similarity. The advent of VisiCalc provided the means to tabulate and present data with summary calculations, and the package became useful and easy to implement. It stayed in this form and was used successfully well into the current IBM PC era.

Two influences spurred the evolution of the statpack into its present format. First was the release of Lotus 1-2-3. Like many VisiCalc power users, I was slow to embrace Lotus but quick to convert when I saw the power available. Second was the desire to convert my software package from Tandy BASIC to IBM format in order to make use of the PC I was using for Lotus.

The first evolutionary step was a decision to work in Pascal or C, rather than just recopy all of the BASIC programs and fit them to Microsoft BASIC. Concurrently, I learned to write Lotus Macros, and the evolution slowly but surely began to drift into a set of templates. These templates performed most of the functions of my original software package, but were written in Lotus macro language and executed as spreadsheets.

Premise

This book is a "how to" text on elementary statistics. It will not cover statistical theory in great depth, but it will explain statistical concepts in a practical manner. For example, the mean and standard deviation are considered in relationship to histograms of real data points rather than the Gaussian equations.

The text is unique because no knowledge of statistics or outside text is required to make use of it. It is best applied, though, by a person with some knowledge of statistics and use of 1-2-3. All listings are provided and can be entered into Lotus release 1 or 2 (some functions are not available in release 1, but these functions are the minority). Sample problems and exercises are divided among scientific, academic, and business applications to make learning the templates easy and the applications broad.

Those who want to minimize typing can purchase the companion diskette providing all of the templates (see order form at the end of book). This set of functional templates is itself a statistical software package. This book is the ideal choice for those who want to use it as a reference manual to the statistics package. The audience for a package of this nature is diverse. In any technical, business, or management environment, students and professionals are bombarded with data in unlimited forms. Those who will make the most of their experience and education will learn to maximize the value of this data.

Whenever data elements are accumulated, by survey, experiment, or any other source, there is but one goal in mind: making a decision. There may be a clear and present need for the information now or a desire to accumulate information for future reference, but the information is always aimed at decision making. Two skills govern the ability to make decisions based on data: analyzing or reducing the data to basic relationships relative to the decision and communicating that data to the decision maker. This book is a tool for improving both of these skills.

Each statistical concept will be presented in terms of the types of variables or data to which it is suited, and logical methods of presenting that data will be included. Whether the user is a student trying to communicate what he or she has learned, an engineer demonstrating developmental test results, or a marketing manager studying test market results, the goals are the same: to understand what has been done and communicate that knowledge to someone else. This book addresses both requirements.

Usefulness

Most 1-2-3 users find, after the initial learning curve, that they can sit down with a data set and extract some useful information fairly quickly. In general, this information includes graphs, data tables, sorts, and memos. Typically, each user develops a few pet tricks and formats that are comfortable. At this point, the system goes flat. Few consider the power of adapting statistical techniques and saving predefined templates for quick application of good methods. Also, presentation-quality output is not often considered. Another memo with a few Lotus graphs begins to look like just what it is — another memo with a few Lotus graphs.

This text will provide a set of proven data reduction techniques. While it can be used as a canned statistics package, it is intended more to provide a stepping stone to true *power* use of Lotus 1-2-3. It will help you explore the hidden corners of the 1-2-3 menus about which you always wondered.

This text will then help you build a set of data tools that will be comfortable for you to use. Not all statistical methods will be covered, but those that are covered will significantly increase your set of methods. The increase in your 1-2-3 abilities will let you expand far beyond the tools presented here to create truly custom templates of your own.

Finally, the presentation tips will allow you to generate custom outputs. These are not sterile, "canned-package" outputs but dynamic and flexible formats. Students can adapt reports to suit instructors and material; professionals can make custom presentations, which can then be transferred to slides or transparencies for overhead projection; and businesspersons can customize reports for customers or publications.

In each template output, three factors are considered:

- Information
- Simplicity
- Ergonomics

Outputs must contain all relevant information, which includes not only results of spreadsheet calculations but the who-how-what-when-where of the data set. Simplicity helps you avoid overwhelming the reader with useless facts and figures. Finally, consider ergonomics; view the output as an entity, and make it easy to look at and understand.

Having completed this work, at the very least I have finally finished what was started seven years ago. It in no way resembles what I started out to do, but it has become twice as useful as I expected. For quick analysis, I generally plug in data values either through the data imput template or directly into the analysis module. Most of the time, I leave the labels blank and work for the outputs. When it comes time for presentations, I build a full data file, and then modify the outputs of the analysis template to suit the occasion. Whether the recipients are supervisors, teachers, or customers, the hallmark of data presentation is mentioning their name or organization in your report. The template structure makes these touches of professionalism simple.

Will Cloake
Ontario, California
January 1988

INTRODUCTION

This book is divided into four major parts. The first is a review of Lotus functions used in this text to develop templates and models. Nearly every user will have a command of some of these techniques, and experienced users will have command over most. At the discretion of the reader, this part can be read, skimmed, or skipped and referred to as reference.

It is presumed that the reader has a working knowledge of Lotus 1-2-3, but that background is not mandatory. Users of the available spreadsheets or neophytes who can learn quickly will have little trouble keeping up with the material presented, particularly if they have access to an experienced user to whom they can go for help. It is also assumed that readers have some experience with statistics, such as an elementary course in engineering or business statistics.

Overall Organization

Part I contains five chapters. Chapter 1 discusses the @ functions and ranges. Chapter 2 covers data base commands. Chapter 3 presents an introduction to macros. Chapter 4 covers methods for getting data to and from disk, and Chapter 5 shows how to create and save graphs in 1-2-3.

Part II contains four chapters that discuss **one-way analysis** methods. These methods are applicable to single data sets where the expected result is the performance of one variable relative to another. One-way methods are the basic statistical methods: mean, variance, deviation, range, etc.

Chapter 6 covers the data file creation template. Chapter 7 addresses the small sample statistics template. Chapter 8 covers the descriptive statistics methods, and Chapter 9 discusses some graphic methods and curve fitting.

Part III consists of three chapters that discuss **two-way analysis** methods. These methods are used for comparing data sets or the performance of two variables against another. Methods are applied to pairs of data sets similar to those in Part II or sets of data taken as pairs.

Chapter 10 includes linear regression and F, T, and Z tests. Those final three tests allow comparison of means and variances of statistical samples. Chapter 11 introduces some graphic techniques to augment the tests in Chapter 10. Finally, Chapter 12 provides some distribution-independent methods.

Part IV includes Chapter 13, a quick reference guide to using the whole package as an integrated unit. If this chapter is scanned prior to reading the rest of the book, the entire text will be easier to understand.

Chapter Organization

Beginning with Part II, the material in each chapter is presented as follows:

- Description of the macro menu structure
- Description of the methods used and types of data suited for each application
- Description of template outputs
- 1-2-3 commands used
- Macro descriptions
- Required range names
- Summary
- Exercises
- Template Listing

This type of chapter organization is intended to walk the user through the development and implementation of each template.

Program Diskette

A companion program diskette is available from the publisher, which contains all of the templates as they are presented in this text (see order form at end of book). Full listings are contained in the book, but by using the companion diskette, users can save a great deal of typing time by working with the statistics package on disk and using this book as a reference manual.

Key Notations

Unless otherwise noted, information enclosed in angle brackets (< >) indicates that you should press a particular key on the keyboard. For example, an instruction to enter

abc<return>

means that you should type in the letters "abc" and then press the Return key (also known as the Enter key).

Notations such as

<Alt/a>

indicate a key combination, which, in this case, is entered by pressing and holding down the Alt key and then simultaneously pressing the letter A key.

Parenthetical expressions or text in standard typeface to the right of commands and key entries are explanations for the reader and are *not* to be entered. For example,

<F5>A18   Cursor at 18

indicates that if you press the F5 key and enter "A18," the cursor will be at column 18. Similarly, the instruction

I (Input)

means to enter the letter I for the Input command.

Information in curly brackets ({ }) represents key commands but should be typed in as text when creating macros. For example, if you type in {End} as part of a macro program listing, then the macro will execute the End key command at that point when the macro is run.

PART I

WARM-UP EXERCISES

Before an intensive look at specific statistical applications for 1-2-3 is begun, intermediate to advanced 1-2-3 functions must be reviewed. Throughout this book, you will build spreadsheet models (or templates) using these techniques. While it is not necessary to fully understand all Lotus 1-2-3 operations to use this book, you will gain the most from the text if you understand as many operations as possible.

Most users of Lotus 1-2-3 (or any software system, for that matter) learn to perform a finite set of tasks using a subset of available commands, which can be the result of education, on-the-job training, or reading a professional text such as this book. While such techniques are adequate preparation for performing many tasks, expanding to new areas can be difficult. For users dealing with numerical data, this text will provide a means for growth. Methods are not restricted to a specific field but resemble those taught in elementary statistics courses in business, marketing, mathematics, psychology, science, engineering, or industrial technology. Examples have been chosen with such diverse applications in mind.

To ensure clarity, sample speadsheets are shown with values, formulas, or both if necessary. Complete listings of all macros are furnished at the end of each chapter. The companion program diskette contains all completed templates and can be used as a statistical software package, with this text serving as a reference manual.

In the five chapters of Part I, you will move rapidly to an advanced level of Lotus use. Chapter titles provide an outline of the material to follow:

Chapter 1	@ Functions and Ranges
Chapter 2	/Data Base Commands
Chapter 3	Macros
Chapter 4	File Operations
Chapter 5	Creating and Saving Graphs

WHY A STANDARD OUTPUT?

Because communication is so important, considerable attention will be given to the way the data analysis is presented. A researcher needs little more than a good statistical calculator and a few graphs to understand the data environment. Without the ability to pass this information to others, however, such work is of little value.

Most Lotus users find themselves playing with data until a sensible format occurs. Eventually, this method is replaced by each user's favorite graph or chart output format. Presentation options in each chapter will help you quickly develop a presentation style suited to your needs.

The data must be complete but not overwhelming. Single-page formats with plenty of white space to make them easier to read are best. Critical bits of information, such as totals and summaries, should be prominently displayed and easy to pick off. Adequate descriptions should be included. Templates that can be used as written are provided, but you are encouraged to incorporate your own style.

The most important considerations are

- who is going to read the report?
- how can I make it understandable?

Remember, your superiors need this information as much as you do, but they may lack your interest and exposure. No one will look for long at an ugly chart or work too hard extracting pieces of information buried in a stack of tables.

Routine reports are much better handled by standard data formats; your results will be more effectively communicated if the reader understands the report format. Finally, if the work is important, use these outputs to enhance a written report, or copy them onto acetate for slide shows.

CHAPTER 1

@ FUNCTIONS AND RANGES

1 The @ Functions and Ranges

Many of the commands and functions provided by Lotus 1-2-3 require a data parameter or parameters to operate, which means that a value, cell location of a value, or a range of cells containing values must be available. Through the use of range statements, these values are passed to mathematical and logical functions. This chapter explores the @ functions—Lotus 1-2-3's most powerful group of direct action commands—and the range statements that provide them with parameter input.

THE @ FUNCTIONS

Many calculations are performed so often, in virtually every environment, that they have been simplified to a group of special functions. Referred to as **at** functions, for the "at" symbol (@) that begins each function, they significantly reduce both typing and complexity. A short example should clarify how and why these functions are used.

Figure 1.1 is a sample spreadsheet excerpt typical of examples presented in this book. In this case, the slash key (/) has been pressed, and the cursor is at cell C3. This example and those that follow are short segments designed to be easily reproduced in a spreadsheet. For a more complete understanding of the concepts, enter them as you follow the text. Most of the exercises are either tutorials or examples to help you enter finished templates.

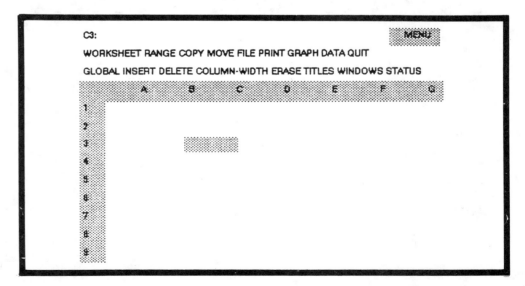

Figure 1.1 Lotus 1-2-3 Main Menu

The @ Functions and Ranges 1

Figure 1.2 is a simple, two-column spreadsheet with four names in column A and corresponding quantities in column B. Cell B6 will contain the total of the four values in column B. Five methods will be presented to accomplish this task. Each of these methods is useful in its own way, and they should all be understood; they have different applications, most of which demonstrate techniques common to all @ functions.

```
B6: (B1+B2+B3+B4)                                          READY


           A            B         C         D         E         F
    1    BOB           19
    2    CAROL         20
    3    TED            5
    4    ALICE         33
    5
    6    TOTAL         77
    7
    8
    9
```

Figure 1.2 Column Summation First Solution

Figure 1.2 demonstrates the first method. For small ranges, this simple mathematical summation — not using an @ function — is the easiest to implement. The equation, shown in the command line, begins with an open parenthesis [(], which enables 1-2-3 to tell the cell reference, B1, for example, from a label beginning with a B. An alternative form is

+B1+B2+B3+B4

1 The @ Functions and Ranges

In either case, the problem is solved using an equation instead of a function. The remainder of the methods use spreadsheet functions. The @ function for addition is @SUM(range). The next four methods each use the sum function with different means of range definition. At the conclusion of this exercise, you should be familiar with how to define and call a range.

RANGE DESCRIPTIONS

The first method of displaying ranges is quickest if the cells containing the range ends are known. Move the cursor to cell B6 and enter the following:

@SUM(B1,B4) <return>

Notice that the first line of the control panel at the top of the screen is slightly different from the one in Figure 1.2. If the entry is correct, the following should be displayed:

B6:   @SUM(B1..B4)

Lotus displays two periods (..) between range entries, even though only one must be entered. As with all these examples, the value displayed in cell B6 is still 77.

The next method uses the techniques of **anchoring**, **painting**, and **fast access** to the range end. Follow these instructions carefully:

1. Keep the pointer at cell B

2. Enter

    @SUM(

3. Using cursor control keys, move the cell pointer to cell B.

    continued...

The @ Functions and Ranges 1

...from previous page

4. Press

    . (a period)

    to "anchor" the range

5. Press

    <End> (numeric keypad key 1)
    <down arrow> (numeric keypad key 2)

The screen should resemble Figure 1.3 with the partially entered formula displayed in the control panel at the top of the screen and the "painted" range highlighted. If something else is displayed, verify that the keypad is not set for numbers (if number mode is active, NUM will be displayed at the bottom of the screen) and carefully reenter the commands.

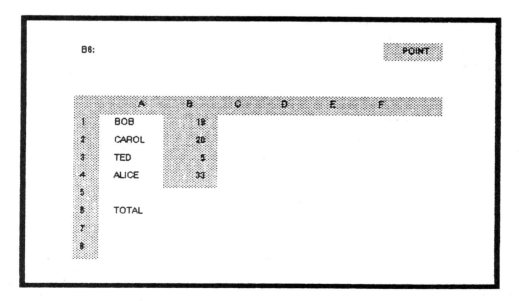

Figure 1.3 Column Summation Range Highlighted

1 The @ Functions and Ranges

To complete the command, type

) <return>

The screen should now look exactly as it did after the second example. This technique is the one most commonly used for range entry because it allows the user to find range ends rather than remember locations, which reduces the potential for errors.

The third technique introduces an extended feature of function range value that allows entry of multiple-range locations, as illustrated by the following instructions:

```
@SUM(B1.B2
,   (a comma)
B4,B3)
<return>
```

In this case, three ranges have been entered and separated by commas. Range 1 is B1-B2, range 2 is the single cell B4, and range 3 is the cell B3. The single cell ranges could have been elsewhere on the spreadsheet, horizontal ranges (i.e., A20..A30), or even constants. The control panel should now contain the following:

B6:   @SUM(B1..B2,B4,B3)

Of course, cell B6 is still 77. To add a constant, as mentioned previously, edit the cell as follows:

1.   Press the Edit key <F2>

2.   Press

   <backspace>   (to remove the closing parenthesis)

   continued...

The @ Functions and Ranges 1

...from previous page

3. Enter

    ,200)

4. Press

    <return>

Cell B6 now contains 277, the sum of the indicated ranges and the numeric constant 200. This application of the range provides readable formulas and is useful for summing multiple columns.

The last range method to be described holds the greatest potential for simplifying complex environments, but it is frequently overlooked. Lotus allows you to give a name to a range that can be used in place of specific locations in @ functions. The procedure is as follows:

1. Move the cursor to B1

2. Type

    /RNC (/ followed by (R)ange (N)ame (C)reate)

3. Type

    COUNTS

4. Press

    <return>
    <down arrow> (3 times)
    <return>

5. Move the cursor to B6

    continued...

1 The @ Functions and Ranges

...from previous page

6. Type

    @SUM(COUNTS)

7. Press

    <return>

Again, 77 is in B6, but the control panel displays the formula with the range name as shown in Figure 1.4. Clearly, this result is overkill for such a simple example, but obviously, in a large spreadsheet, reference to ranges by name can be an effective simplification tool.

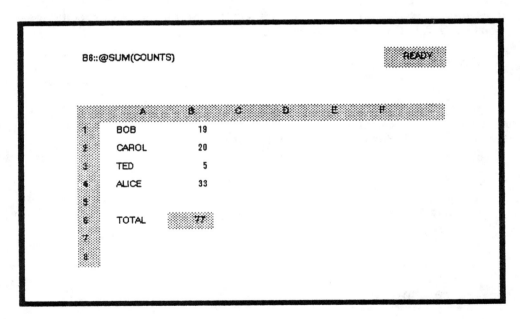

Figure 1.4 Column Summation Named Range

The @ Functions and Ranges   1

After you closely examine ranges (the arguments for @ functions), studying the functions becomes easier. There are 49 unique functions in Lotus release 1A and an additional 29 in release 2. These functions, with definitions, are listed in the two tables at the end of this chapter. They are grouped by category as follows:

- Mathematical
- Special
- Financial
- Statistical
- Logical
- Data Base Statistical
- Date

Release 2 includes two additional groups:

- Time
- String

In most cases, these functions can be thought of as formulas or numbers because they have a value and can be used in other formulas. The previous example could have been written with cell B6 as

@SUM(@SUM(B1,B3),(B2,B4))

The value in cell B6 (still 77) is the sum of the boys plus the sum of the girls. Simply put, these functions can be used anywhere a number would be placed.

1 The @ Functions and Ranges

EXAMPLE PROBLEM

Enter the data values that follow into the first eight cells in column A. Starting with cell C1, display the following calculations based on the data values given:

- minimum
- maximum
- average
- count
- range

Values

33
-100
63.33
27
@sin(@pi/6)
@mod(10,3)
0
1.2e4

Solution

Enter the eight data values into column A *exactly* as shown. Move the cursor to C1 and enter the following. Remember, information in angle brackets (< >) indicates a key to be pressed.

```
/rncdata<return>A1.A8<return>
@min(data)<down arrow>
@max(data)<down arrow>
@avg(data)<down arrow>
@count(data)<down arrow>
@max(data) - @min(data)<return>
```

Figure 1.5 shows the result of these actions. Several interesting things have happened. In cell A5, the sin of 30 degrees (pi/6 radians) is displayed as 0.5. A6 displays 1 as the remainder of 10/3 (ten modulo three). A8 contains 12000, which is the equivalent of the exponential term 1.2e4.

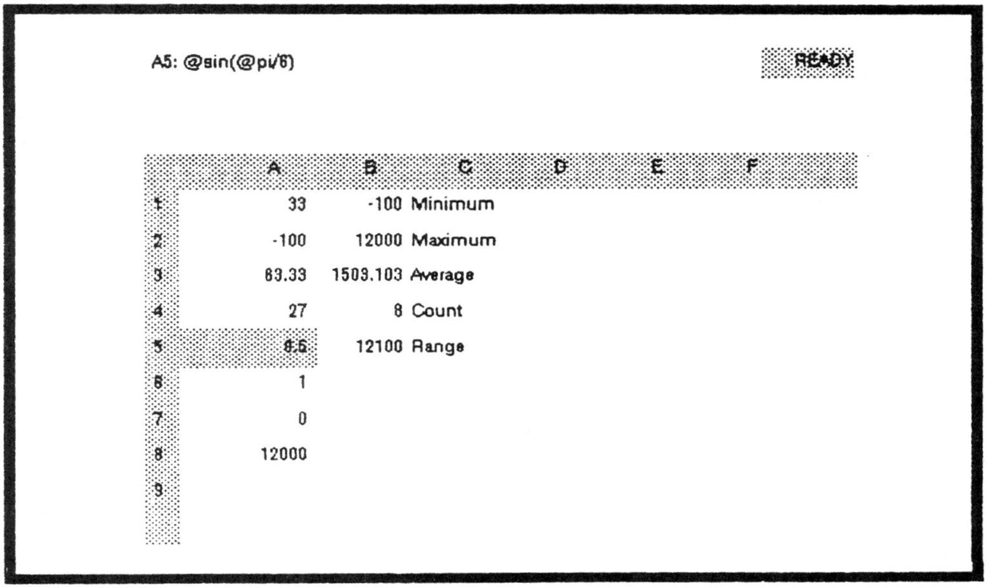

```
A5: @sin(@pi/6)                                              READY

         A           B          C          D        E       F
 1      33         -100    Minimum
 2    -100        12000    Maximum
 3   83.33     1503.103    Average
 4      27            8    Count
 5     0.5        12100    Range
 6       1
 7       0
 8   12000
 9
```

Figure 1.5 Examples of @ Functions

Note that in A6 and C5, @ functions were used in place of numbers. An @ function is a value, so it can be used in place of a number just as a cell name can. The function is evaluated, and the result is used in the equation.

1 The @ Functions and Ranges

SUMMARY

- Use of templates and standardized data formats reduces the chance for error and increases readability of the data presented.

- The @ functions generate values that can be used as numbers in equations.

- A variety of methods exist for presenting ranges in functions, each with advantages.

- Use of the "anchor-paint" technique for range entry is simplest and minimizes errors.

- Ranges can be referred to by name.

EXERCISES

Case Study One: Widget Sales

Assume the data in Figure 1.3 are one month sales, in thousands, for the west coast office of Widgets, Inc. Reproduce that table and add the following information for the east coast office:

Tom	21
Dick	11
Harry	51
Jane	22

Enter the information for each office, including total sales, in two columns with headings indicating east/west. Use this table to solve the four problems that follow.

1. Below the column totals, display minimum, average, and maximum for each office.

continued...

The @ Functions and Ranges 1

...from previous page

2. Label cell A12 as "SUMMARY:" Starting in cell A13, provide a table with the composite minimum, average, maximum, and count of salespersons and total sales.

3. Expand the table to include sales information for men and women in separate columns.

4. The following figures represent the sales information for the next month. Repeat the exercise for this information. Note that Tom has left the East Coast office:

Bob	27
Dick	33
Carol	31
Harry	44
Ted	19
Jane	19
Alice	34

Answers

Figure 1.6 shows the spreadsheet prior to addition of the information in the exercises. Note that the headings WEST and EAST are offset so that each covers both of the respective columns. The contents of B7 (and D7) are shown as follows:

@SUM(B2..B5)

There are several ways to define the total, but Figure 1.6 used the sum function.

1 The @ Functions and Ranges

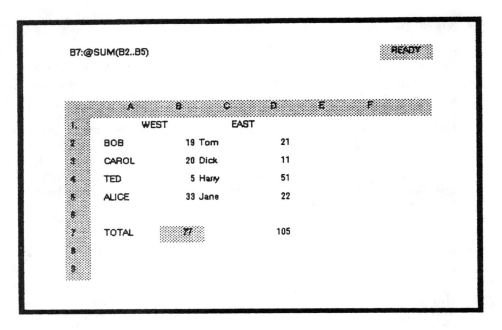

Figure 1.6 Data Table for Chapter 1 Exercises

1. The six statements that follow will provide the labels and results for the information in column B. In each case, move the cursor to the indicated cell and enter the information shown.

   ```
   A8:  'Minimum
   A9:  'Average
   A10: 'Maximum
   B8:  @MIN(B2..B5)
   B9:  @AVG(B2..B5)
   B10: @MAX(B2..B5)
   ```

   The calculations for the D column can be entered with the following commands:

   /cb8.b10<return>d8<return>   / Copy

The @ Functions and Ranges 1

The information in the next three lines will be generated by the following command sequence:

```
D8:  @MIN(D2..D5)
D9:  @AVG(D2..D5)
D10: @MAX(D2..D5)
```

The addresses D2-D5 are entered in the range portion of the functions because the copy statement inherently uses relative addressing unless otherwise indicated.

2. The following statments will add the composite calculations and labels:

```
A12: 'SUMMARY
A13: 'Minimum
A14: 'Average
A15: 'Maximum
A16: 'Count
A17: 'Total

B13: @MIN(B2..B5,D2..D5)
B14: @AVG(B2..B5,D2..D5)
B15: @MAX(B2..B5,D2..D5)
B16: @COUNT(B2..B5,D2..D5)
B17: @SUM(B2..B5,D2..D5)
```

Each of the statements in this group has a multiple range, which consists of two ranges separated by a comma.

1 The @ Functions and Ranges

3. The final two groups of statements provide the two columns of information for men's and women's sales:

```
C12: 'Men
C13: @MIN(B2,B4,D2..D4)
C14: @AVG(B2,B4,D2..D4)
C15: @MAX(B2,B4,D2..D4)
C16: @COUNT(B2,B4,D2..D4)
C17: @SUM(B2,B4,D2..D4)

D12: 'Women
D13: @MIN(B3,B5,D5)
D14: @AVG(B3,B5,D5)
D15: @MAX(B3,B5,D5)
D16: @COUNT(B3,B5,D5)
D17: @SUM(B3,B5,D5)
```

Ranges in the men's group are a mixture of single-cell ranges and a multiple-cell range. In the women's group, each range has three distinct cells (single-cell ranges). Figure 1.7 shows the completed spreadsheet.

The @ Functions and Ranges 1

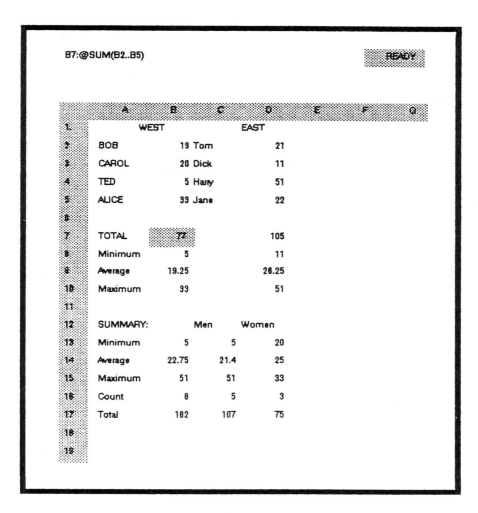

Figure 1.7 Solution to Exercise 1.3

4. The fourth exercise is trivial because the new data values can be entered in place of the old, and the spreadsheet will automatically update all of the calculations. The only trick is eliminating Tom from the display rather than entering a blank for his sales. Enter the following:

/mc4.d5<return>c3<return>   / Move

21

1 The @ Functions and Ranges

When this command sequence is complete, substitute Dick's name for Tom's in cell C2. If the range had been simply erased, the totals and counts would have been correct, but the top cell in the display would have been left blank. If the move statement had been as follows,

/mc3.d5<return>c2<return>   / Move

(which may have seemed the obvious choice), the range statements all would be erased and each would indicate "err."

The important point to remember is that a range may be expanded or contracted as long as neither end is affected. Figure 1.8 shows the final solution.

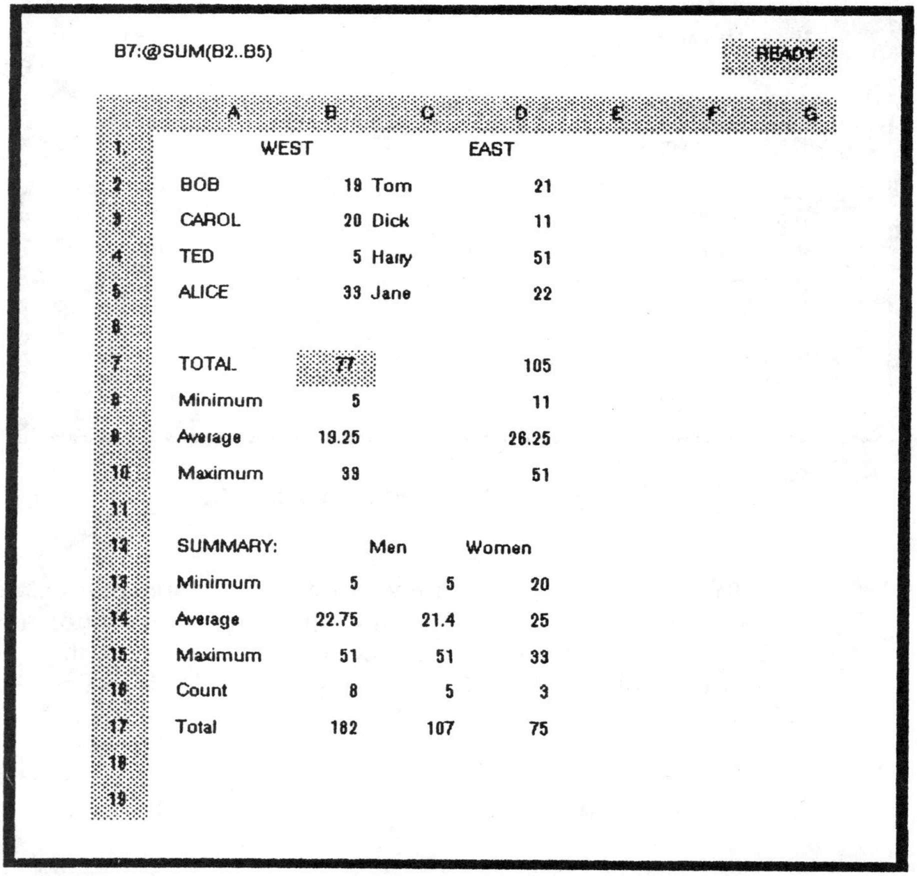

```
B7:@SUM(B2..B5)                                              READY

          A           B           C           D           E           F           G
  1.                WEST                    EAST
  2.    BOB          19   Tom              21
  3.    CAROL        20   Dick             11
  4.    TED           5   Harry            51
  5.    ALICE        33   Jane             22
  6.
  7.    TOTAL        77                   105
  8.    Minimum       5                    11
  9.    Average    19.25                 26.25
 10.    Maximum      33                    51
 11.
 12.    SUMMARY:         Men         Women
 13.    Minimum       5           5          20
 14.    Average   22.75        21.4          25
 15.    Maximum      51          51          33
 16.    Count         8           5           3
 17.    Total       182         107          75
 18.
 19.
```

Figure 1.8 Solution to Exercise 1.4

The @ Functions and Ranges 1

LIST OF FUNCTIONS

The list in Table 1.1 is an introduction to the functions needed to build the models in this text. They are described here for reference only and will be fully defined as they are used later in the book.

@MOD(x,y)	Remainder from x/y
@ROUND(x,digits)	x rounded to (digits) digits
@INT(x)	x truncated
@SQRT(x)	Square root of x
@IF(x,true,false)	Conditional branch
@CHOOSE(t,v0,..,vn)	Selects value from list
@(h,v)LOOKUP	Picks value from lookup table
@COUNT(list)	Number of elements in range
@SUM(list)	Sum of values in range
@AVG(list)	Average of values in list
@MIN(list)	Smallest value in list
@MAX(list)	Largest value in list
@STD(list)	Standard deviation of list
@VAR(list)	Variance of list
@DCOUNT	Field count
@DSUM	Field sum
@DAVG	Field average
@DMIN	Field minimum
@DMAX	Field maximum
@DSTD	Field standard deviation
@DVAR	Field variance

Table 1.1 Lotus Release 1A Functions

1 The @ Functions and Ranges

Additionally, the following functions are for Lotus release 2:

@RAND	Generates random number
@ROUND(x,n)	Rounds x to n digits
@@(cell)	Contents of cell (cell)
@COLS(range)	Number of columns in range
@INDEX(range, column,row)	Value of cell at Row,Col
@ROWS(range)	Number of rows in range

Table 1.2   Lotus Release 2.01 Functions

CHAPTER 2

/DATA BASE COMMANDS

2 The /Data Base Commands

Lotus 1-2-3 @ functions operate directly on a range of cells that are all treated identically. These methods are used, at least in part, by almost all 1-2-3 users. Data base commands operate over a range of cells in which each column is treated as a location for particular data types. Each row in the range contains elements that relate to a single item in the data base. Therefore, the data in a given column are similar in form, while the data in a particular row are representative of each data type.

A **data base** is a collection of groups of data. Each group, usually called a **record**, is broken into one or more subgroups called **fields**. Data base management systems (**DBMSs**) are software packages designed to store, access, calculate, and report on data groups. Information is entered into the fields that are grouped and stored as records on the disk. There are many DBMS programs available, but dBASE from Ashton-Tate is the most widely used.

In 1-2-3, a data base is a collection of cells. A record might be a row with fields of information in different columns. By this definition, the examples in Chapter 1 are small data bases.

Many types of information are gathered for storage and reporting. For the purposes of this book, two relevant types are **data sets** and **informational records**. Data sets, such as the heights of American males or records of a corporation's sales, might contain large numbers of records with small fields. In speadsheet terminology, this means many rows containing a few narrow columns.

Informational records usually contain a few wide columns or multiple rows in the individual records. Medical records in your doctor's office are a good example of informational records. Checkbooks and telephone lists fall somewhere in between the two categories.

Several guidlines can be applied when deciding which system to use. There is no correct answer; both systems can be applied to most applications. Common sense dictates that you use the system most suitable to your needs. Few people use both systems. The following rules of thumb may help:

- For short string and numerical elements, use 1-2-3.
- For long text elements, use a DBMS.
- Reports with long and varied text are suited more to a DBMS.
- Standardized report formats with fill-in numbers or short text elements are easier with 1-2-3.
- A DBMS is more convenient when entire records must be retrieved and displayed as is.
- 1-2-3 makes recalculation and "what if...?" analysis easier.
- If the primary interest is mathematical, use a spreadsheet; if it is informational or if input/output format is critical, try a DBMS.

Although data base management systems and 1-2-3 both have strengths, each excels in different areas. For example, a library card file is much easier to work with on a DBMS, but working with statistics on data sets is easier with 1-2-3.

DATA BASE COMMANDS

Lotus 1-2-3 release 2 offers eight data base commands that are executed under the /Data menu:

- Fill
- Table
- Sort
- Query
- Distribution
- Matrix
- Regression
- Parse

2 The /Data Base Commands

Figure 2.1 illustrates how the commands appear on a spreadsheet after a /(D)ata is executed.

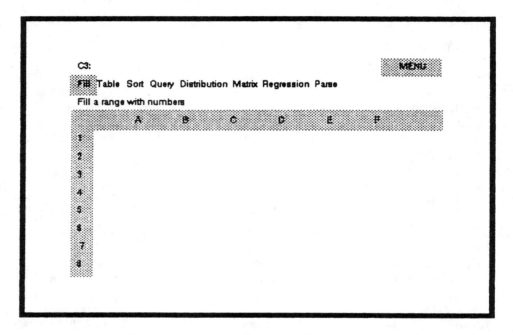

Figure 2.1 1-2-3 Data Base Command Menu

Fill

The Fill command allows any range to be filled with data values. Filled data ranges are then available for mathematical operations as were defined in Chapter 1. Values can be entered as integers or real numbers; ascending or descending; and in single-row, column, or table format.

The /Data Base Commands 2

The first useful application of this command is filling a column with integers to act as row labels. The following sequence fills the first 21 cells of column A with the cells' row numbers:

User keystrokes	1-2-3 Response
HOME	Cursor to cell A1
/<D>ata<F>ill	Enter Fill range: A1
.<PgDn><return> **	Start: 0
1<return>	Step: 1
<return>	Stop: 8095
<return>	

** Press <ESC> before the period (.) if Fill range exists.

Figure 2.2 illustrates the spreadsheet after these actions are taken. In the third line, the response is a period to anchor the range, followed by a <PgDn> (equivalent to 20 <down arrows>), and a <return> to close the range. The fourth line changes the starting value of the range to be filled from zero to one. The next two lines provide the default values of increment (equal to 1) and stop (at 2047). The **increment** can be any number (positive or negative), and need not be whole. The **stop value** is one of two means to limit the range. If a value of ten is selected, only the first ten rows would have been numbered. The last cell in a Lotus release 1A spreadsheet is 2047. In release 2, the last cell is 8095. The second limit on the range's end is the **data range** selected in lines 2 and 3.

2 The /Data Base Commands

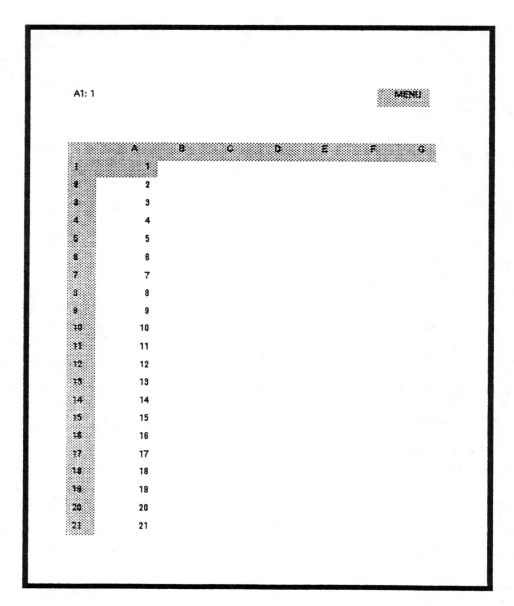

Figure 2.2 1-2-3 Data Fill Command Example

The /Data Base Commands **2**

One final example will demonstrate the flexibility of this command. The following sequence will fill a data table with negative real numbers:

User Keystrokes	1-2-3 Response
HOME	Cursor to cell A1
/<D>ata<F>ill	Enter Fill range: A1
.<down><down><down> **	A1-A4 painted
<right><right><right>	A1-D4 painted
<return>	Start: 0
5.7<return>	Step: 1
-3.3<return>	Stop: 2047
-35<return>	

** Press <ESC> before period (.) if Fill range exists

Figure 2.3 displays the created data table. The <down> and <right> commands represent executing the <down arrow> and <right arrow> key commands. Note that cell A1 has the starting value of 5.7. Data increases by the -3.3 step value going down the column to -4.2. Cell B1 is the next step, and the last value occurs in cell D2. The table is not completely filled because a stop value of -35 was set. When Data Fill is used, at least one value is printed, even if no values in the range are lower than the stop value. The last value printed depends on the Lotus release used (1.1 prints one value after the stop, -37.2; 2.01 stops at -33.9, the last value before -35).

```
A1: 5.7                                                    MENU

              A         B         C         D      E     F     G
     1       5.7       -7.5     -20.7     -33.9
     2                 2.4     -10.8      -24
     3                -0.9     -14.1     -27.3
     4                -4.2     -17.4     -30.6
     5
     6
     7
```

Figure 2.3 Second Data Fill Example

2 The /Data Base Commands

Data Sort

The Sort function is powerful and fun to use. With this command, a group of cells can be reorganized into an alphabetical sequence. Figure 2.4 shows the sample data base used to investigate the sort commands.

```
A1:                                                    READY

         A         B          C            D     E     F     G
    1              CITY       STATE        TEMPERATURE
    2         1    Anaheim    California   65
    3         2    Ontario    California   76
    4         3    Kalamazoo  Michigan     23
    5         4    Albany     New York     13
    6         5    Buffalo    New York     18
    7         6    New York   New York     19
    8         7    Pittsburgh Pennsylvania 11
    9         8    Portland   Oregon       37
   10         9    Boston     Massachusetts 12
   11        10    Cleveland  Ohio         14
```

Figure 2.4 Sample Data Base for Sort Studies

The data chosen represents temperatures on a typical mid-winter day in various places across the country. There are four columns in the present data base. These columns are referred to as **fields**. The first field, filled with integers, contains **record** numbers. The second field contains cities; the third field contains states, and the final field contains temperatures. The only limits to the number of fields are the number of columns and the RAM available. Each row in the data base is a record. A record need not have data in every field, but all information in a row pertains to a single record. Generally, no outside information is considered part of the data base.

The /Data Base Commands 2

The first sort to be illustrated is by temperature and can be accomplished with the following commands:

<u>User Keystrokes</u>

HOME<down>
/<D>ata<S>ort
Data-Range
.<down><down><down><down>
<down><down><down><down>
<down><right><right><right>
<return>
Primary-Key

<right><right><right>.
<down><down><down>
<down><down><down>
<down><down><down> <return>
A<return>
G<return>

<u>1-2-3 Response</u>

Cursor to A2
Sort Menu
Enter Data range: A1

A2-D11 painted
Sort menu
Enter Primary sort key
address: A2

Enter Sort order (A or D):D
Sort menu

The sorted spreadsheet is shown in Figure 2.5. All four columns are included because they are all in the data range defined in lines 3-7 of the example. It is sometimes a good idea to sort a column of sequential numbers along with a data base so the numbers can be resorted back into their original order (see Figure 2.6).

2 The /Data Base Commands

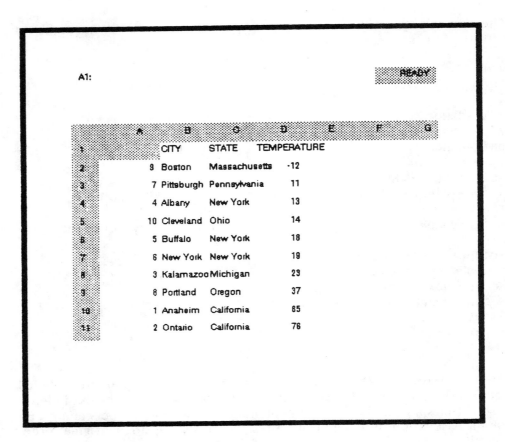

Figure 2.5 Data Base Sorted on Temperature

One additional sort will be illustrated here. The following statements cause the same data base to be sorted by state. Since three cities in New York are listed, the Secondary-Key command is used to alphabetically sort the cities in each state.

The /Data Base Commands 2

User Keystrokes	1-2-3 Response

Reset

Previous settings cleared
Sort Menu
Menu cursor on Reset

Data-Range.\<right>\<right>
\<right>\<down>\<down>\<down>
\<down>\<down>\<down>\<down>
\<down>\<down>\<return>

Sort Menu
Menu cursor on Data-Range
Enter Primary sort key
address: A2

Primary-Key

\<right>\<right>.
\<down>\<down>\<down>\<down>
\<down>\<down>\<down>\<down>
\<down>\<return>
A\<return>

Enter Sort order (A or D):A
Sort Menu
Menu cursor on Primary-Key
Enter Secondary sort key
address: A2

Secondary-Key

\<right>.\<down>\<down>
\<down>\<down>\<down>\<down>
\<down>\<down>\<down>\<return>
D\<return>

Enter Sort order (A or D):D
Sort Menu
Menu cursor on Secondary-Key

Go

The data have again been sorted, this time leaving the states in alphabetical order, with California and New York cities subsorted in reverse alphabetical order.

2 The /Data Base Commands

```
A1:                                                    READY

            A         B         C          D       E    F    G
    1                CITY      STATE      TEMPERATURE
    2             2  Ontario   California    76
    3             1  Anaheim   California    65
    4             9  Boston    Massachuset   12
    5             3  Kalamazoo Michigan      23
    6             6  New York  New York      19
    7             5  Buffalo   New York      18
    8             4  Albany    New York      19
    9            10  Cleveland Ohio          14
   10             8  Portland  Oregon        37
   11             7  Pittsburgh Pennsylvania 11
```

Figure 2.6 Data Base Sorted on Two Levels

DATA QUERY OPERATIONS

Most computer users have entered a list in Lotus 1-2-3. In many cases, this list is an address book or company phone directory. Activity logs, inventories, and purchasing records are other examples of user data bases you may already have tried. These lists are easy to enter, convenient to update, and more convenient to use than manual logs kept on paper.

The / Data Query commands allow any user with a Lotus data base to scan all the records for key words or numbers. If the data base is a phone list, you can Query it for all persons named Smith, everyone living in Toledo, or all phone numbers in a selected area code. One or more of the fields in the data base can be scanned based on user-selected criteria.

At least two ranges are required to Query a data base. Unlike most Lotus command structures, in which only the data are included in a range, the input range for a data base must include the column headings. By default, when Query operations are performed, these headings become the field names. The criterion range includes field names and test conditions necessary to perform a search. Finding all Smiths in a phone list requires an input range with a field named "Last_Name" and a criterion range containing the field name and range, with *exactly* the same spelling. A space on the end of either entry changes its value (as it must), so make these entries carefully.

/Data Query Find

Another example should simplify this discussion. Figure 2.5 contains the current data base sorted by state and city. The following commands will set the input and criterion ranges to allow searching of the data base. The result is similar in action and response to setting the Data Sort range and Primary-Key range in the previous section:

User Keystrokes

<HOME> /DQ

I(input)
.<right><right><right>
<down><down><down><down>
<down><down><down
<down><down><down><return>

C (criterion)
<down><down><down><down
<down><down><down><down>
<down><down><down><down>
<down><right><right>
.<right><down>
<return>

1-2-3 Response

Data Query Menu
Menu cursor on input
Enter Input Range: A1

Input range entered
Data Query Menu
Menu Cursor on Input
Enter Criterion range: A1

Data Query menu
Cursor on Criterion

2 The /Data Base Commands

The range entry is completed in preparation for searching. All that remains is for you to enter the information for which you are searching. Recall that a range was prepared from cells C14-D15 for this purpose. Quit Data mode and enter the following criteria:

Cell	Enter
D14	State
C14	Temperature
D15	New York
C15	13
A14	CRITERION:

Return to Query (/DQ) and execute the (F)ind command. Several results occur: the mode indicator (upper right-hand corner) registers FIND, the cursor disappears, and the "Albany, New York" record is highlighted completely. Figure 2.7 shows the current spreadsheet. If a beep sounds and none of the above happens, check the following:

1. Make sure the criterion range is identical to the range names in Input. Capitalization or extra spaces after the names are treated as differences.

2. Verify that the input range includes the headings and all four columns.

3. Double-check the criterion range (C14..D15).

The /Data Base Commands 2

```
A1:                                                          READY

           A          B          C            D        E       F       G
    1                 CITY       STATE        TEMPERATURE
    2              2  Ontario    California      76
    3              1  Anaheim    California      65
    4              9  Boston     Massachusetts  -12
    5              3  Kalamazoo  Michigan        23
    6              6  New York   New York        19
    7              5  Buffalo    New York        18
    8              4  Albany     New York        13
    9             10  Cleveland  Ohio            14
   10              8  Portland   Oregon          37
   11              7  Pittsburgh Pennsylvania    11
   12
   13
   14     CRITERION: TEMPERATURE      STATE
   15                  13             New York
   16
```

Figure 2.7 Data Query Find Example One

Given the data base established earlier, you were able to set up a range on the spreadsheet where you defined the information to be retrieved from it. In this example, you wanted to list all cities in New York State that had a temperature of 13 degrees. Much work is required for this setup, but you can re-search the data base frequently with a single keystroke. After you make a criterion change, the Query key will initiate a search for records meeting the new standards.

2 The /Data Base Commands

Quit the Query mode (type <return>Q) and change the temperature indicated in cell C15 to 18 (for 18 degrees). Press the F7 function key. The last query command executed is repeated (/DQF) and finds "Buffalo, New York." From here, pressing <return> takes you to the Lotus ready mode. After the temperature criterion is changed from 13 to 18, only a single keystroke is needed to search the data base again. Change D15 to Ohio and press <F7>. The beep(s) indicates "no record found" because no city in Ohio had a temperature of 18 degrees. Finally, change the temperature to 14 degrees and press <F7>. "Cleveland" is found, and its record is highlighted. The data base is sorted again with a single keystroke each time. Once you have set the input and criterion ranges, you can repeat these operations as often as desired for any values. Don't forget to save the data file after changes are made to input and criterion ranges.

If a single value is used as a criterion, multiple searches are required to find groups of data. To find all cities with freezing temperatures, you would have to repeat the search for 32 degrees, 31, 20, etc., on down to the lowest temperature in the data base.

To alleviate this problem, you can use formulas as criteria. Use /Range Erase to clear cell C15, and put the equation +D2<=32 in cell D15. Now, any city in the United States where the temperature was at or below the freezing level meets the criterion. When the F7 key is pressed, "Boston, Massachusetts" is found. As you press the down arrow key, each record containing a freezing temperature will in turn be highlighted. Anaheim, California and Portland, Oregon will be highlighted whether they had freezing temperatures or not because the Home and End keys position the highlight at the top and bottom records in the data base, whether or not they meet the criterion. If the spreadsheet is wider than 76 columns, the left and right arrow keys scroll the small cursor, allowing you to view cells.

Saving the spreadsheet saves all graph and data settings. The next time this spreadsheet is loaded, pressing the F7 key will again highlight the records satisfying the last criterion entered prior to the /FS command.

Up to this point, all the criteria chosen were related by a logical AND. The data base was searched for temperatures found in New York *and* equal to a given value. This AND occurs when the criterion values for two or more elements are on the same criterion line as the Range. When "New York" and "13" were in cells C15 and D15 in the previous example, "Albany" was found in the data base. If two (or more) values are offset, an OR relationship is used to sort the data base. The following commands will establish a search for temperatures in California cities *or* places below 15 degrees.

The /Data Base Commands 2

```
F5 (goto) D15
/M<return><down><return>
<down>California<return>
<left><up>+d2=32
F7
```

Note that the Move command automatically adjusts the criterion range. The cursor finds "Anaheim" and "Ontario" because they are in California and finds "Boston," "Albany," "Cleveland," and "Pittsburgh" because they are below 15 degrees.

/Data Query eXtract

Now that you can search a data base for specific information, the next logical step is to extract those bits of data into a separate table of their own. The eXtract command serves this purpose (see Figure 2.8).

```
A1:                                                       READY

           A       B          C           D       E    F    G
  1                CITY       STATE       TEMPERATURE
  2             2  Ontario    California       76
  3             1  Anaheim    California       65
  4             9  Boston     Massachusetts   -12
  5             3  Kalamazoo  Michigan         23
  6             6  New York   New York         19
  7             5  Buffalo    New York         18
  8             4  Albany     New York         13
  9            10  Cleveland  Ohio             14
 10             8  Portland   Oregon           37
 11             7  Pittsburgh Pennsylvania     11
 12
 13
 14      CRITERION            TEMPERAT STATE
 15                               0
 16                                    California
 17
 18      TEMPERATURE          CITY     STATE
 19
```

Figure 2.8 Data Query Output Range Selected

41

2 The /Data Base Commands

Data eXtract

As with the ranges selected in Data Find, eXtract requires an output range. Figure 2.8 illustrates how an extract range is added to the previous example. The following commands will generate the added extract range:

```
<F5>A18<return>                              Cursor at A18
Temperature<right>
City<right>
State<left><left>
/DQO (Data Query Output)                     Enter output range: A18..
.<right><right><right>
<down><down><down><down>
<down><down><down><down>
<return>                                     Range painted
                                             Query Menu
                                             Menu cursor on Output
```

Entering eXtract now locates all records that satisfy the criteria. Instead of being highlighted, they are transferred to the output range and simultaneously displayed. The records become a permanent part of the spreadsheet, just as if they had been entered from the keyboard.

Figure 2.9 shows the completed spreadsheet. As with Figure 2.7, since the criteria have not changed, "Anaheim," "Ontario," "Boston," "Albany," "Cleveland", and "Pittsburgh" are found. These records are transferred to the output range. Note that the headings from the original range are again used to point to the range for output. These headings must be exactly identical to the previous headings, except for the order; they have been rearranged in the output range. It was not necessary to print all of the input ranges in the output section; only those with headings on the columns show up, and the rest are ignored.

The /Data Base Commands 2

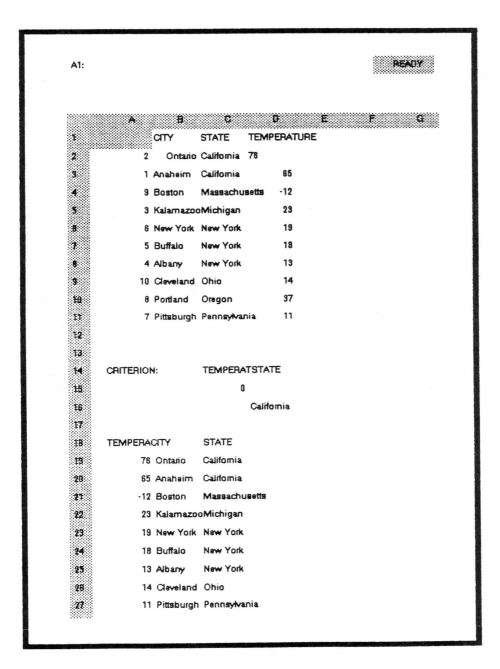

Figure 2.9 Data Query Extract Executed

2 The /Data Base Commands

Additional Data Commands

The **Unique** command is identical in form and function to the eXtract command, but it does not transfer duplicate entries. It is especially useful in spreadsheets containing blanks, as only one blank row will be printed.

The data **Table** command allows automation of "what if...?" tables for forecasting, but these tables are not relevant for the applications developed in this book.

Data **Parse** is a powerful command that allows import of data from a DBMS or custom software application. Literally, "parse" means to separate into parts for individual analysis. In Lotus terminology, it is a method for breaking a single column of cells into several columns of cells over a range in the spreadsheet.

Data to be parsed are generally imported from a DBMS or application program. When data are imported into 1-2-3, you have two choices: data can be treated as text or numbers. In numbers mode, all text in the file is ignored. In text mode, each row is imported into a single column as one long string of ASCII characters.

The limitation of the Parse command is that all rows of entries are parsed at the same positions as the first row. If data are imported from a DBMS, this is no problem; fields are always the same width. When data are imported from applications or custom programs, prior planning is required to avoid difficulty in performing the parse. Detailed study of this command can be found in Lotus manuals such as *The Power of: Running 1-2-3* (Robert E. Williams, MIS:Press, 1988).

Regression (and discussion of the /Data Regression command) will be covered in Part III of this book.

SUMMARY

- A data base is a collection of cells usually containing information about a specific topic.

- Lotus can be used interchangeably with DBMS systems such as dBASE III but is best suited to data bases with short text or numerical elements requiring frequent recalculation.

- /Data Fill allows entry of columns or tables of uniformly spaced real numbers.

- /Data Sort allows a two-level table sort.

- /Data Query allows a search of a data base for records meeting one or more specified criteria.

- A data base is a set of records broken into fields of data.

EXERCISES

The exercises in this chapter are based on the case study that follows. Create a spreadsheet for these exercises based on the following information, and save it for use in later chapters.

Case Study Two: Multiple-Office Sales Department

XYZ Incorporated has three sales offices: East, Midwest, and West. Twenty-four salespeople are spread among the three offices. The following table provides the name, location, sales for this year to date and last year, sales position, and seniority level (tenure rank) for each salesperson. Create a data base from this information, and save it in a disk file. All the exercises in this chapter are based on the following information.

2 The /Data Base Commands

Name		Sls	Sales (units)			Ten
Last	First	Off	YTD	Prev	Pos	Rnk
Nikodym	Paul	E	217	205	M	1
Sager	Justin	E	215	175	Sv	2
Dhanoa	Harpreet	E	163	150	I	3
Sathoff	Jamey	E	199	214	O	4
Sanchez	Abraham	W	312	270	Sv	5
Reyes	Christie	W	52	0	I	20
Matlock	Debbie	W	166	111	O	6
Fuller	Kelley	W	102	88	O	7
Manning	Kimberly	W	102	88	I	8
Tran	Kim	W	34	10	I	9
Dyer	Kenny	M	198	121	O	10
DeStuben	Thomas	M	77	77	I	11
Atilano	Mark	M	189	47	O	12
Armijo	Christina	M	234	112	Sv	13
Brake	Brian	M	23	25	I	14
Burrows	Holly	M	19	0	I	15
Green	Lee	M	47	0	I	16
Garcia	Gilbert	E	111	0	I	17
Lopez	Mary	M	38	0	I	18
Floyd	Erbin	M	74	0	I	19
Guittierez	Brenda	W	74	0	I	21
Johnson	Eddie	M	22	0	I	22
Cochavats	John	E	5	0	I	23
Zitterkoph	Brenda	W	18	0	I	24

Table 2.1 XYZ Inc. Sales Information

Legend:

Rec	Record Number
Sls Off	Sales Office
E	East
M	Midwest
W	West
Sales YTD	Year-to-date Sales
Sales Prev	Previous Year Sales

continued...

...from previous page

Pos	Sales Position
M	Manager
Sv	Supervisor
I	Inside Sales
O	Outside Sales
Ten Rnk	Tenure Rank (seniority level)

Exercises

1. Sort the data base alphabetically and print the results. Re-sort the data by year-to-date sales, subsorted by last year's sales, both in descending order, and print the result. Sort the file on sales office in descending order with last names alphabetized, and print the result. Finally, restore the data to its original order.

2. What record contains the top salesperson based on year-to-date sales? What is in the Pos field of this record, and what field contains the string "E"?

3. Set up a criterion table containing the sales office. Execute a Find operation with an input range that includes all current records and fields and has the criterion table as the criterion range. Put a value of M in the criterion table and execute a find.

4. Set up an output range and extract the staff of each sales office, printing the results. **Note:** This exercise requires three separate operations.

5. Sort the data base again based on sales office. Extract the entire data range to another spreadsheet. Load the extracted spreadsheet; then, create a summary table that includes minimum, average, and maximum sales by office and grand totals for YTD and previous year's sales totals. Print the results.

2 The /Data Base Commands

Answers

An infinite number of correct answers to problems of this type exist, but some key points in data base operation must be considered. Each line in the spreadsheet is a record that must contain all of the related pieces of information. If three tables were set up—one for each office—then three effective data bases were made instead of one. Figure 2.10 shows the first few lines of this data base. The answers that follow are based on this setup.

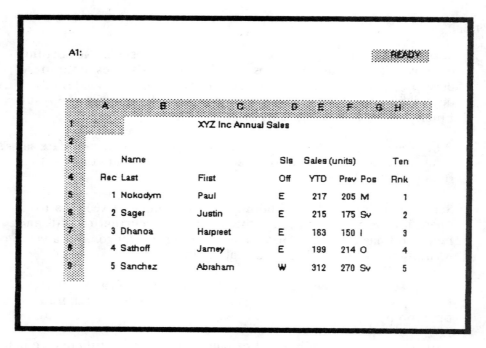

Figure 2.10 Data for Exercises

1. Three or four sorts are required, depending on the how you restore the data. First is the alphabetic sort:

   /dsrda5.h28<return>p5.b28<return>a<return>
   sc5.c28<return>a<return>g

The /Data Base Commands 2

In the first line, /dsr (data sort reset) resets all existing sort ranges and parameters. It is not necessary to reset these parameters every time sort is used, but sometimes it is easier to reset them than to move the ranges. The data range is A5-H28 (set by da5.h28<return>). Primary sort is on last name, acsending (pb5.b28<return>a<return>). The second line sets the secondary key to first name and executes the sort. Since no two sales people had the same last name, it is actually unnecessary to set the secondary key.

The second sort is two levels, both in descending order:

/dspe5.e28<return>d<return>sf5.f28<return>d<return>g

This time, no reset was executed after /ds, so the data range did not need to be entered. Remember that if you are not certain of the exact end points of the ranges 1-2-3 allows, use of the cursor keys to find them. After you press /dsp (data sort primary-key) — the cursor keys allow you to move to the first cell in the range (E5 here) — press "." and move to the last cell (down arrow keys or the end-down sequence) and press <return> again. The third sort requires descending order by sales office, subsorted alphabetically. Figure 2.11 is a copy of the printout of the data base after this third sort.

The final requirement is to restore the file to its original order. It can be sorted on record number to reorder the records, or you can simply load the file from disk again using file retrieve (/FR).

2. If the data base is sorted on YTD sales, the top record is the top salesperson, which is record 5, "Sanchez, Abraham." The position field contains "Sv" for supervisor, and "E" for East is in the sales office field.

3. The criterion field can be set up with the following entries:

   C52: 'Criterion:
   C53: 'Off
   C54: 'M

49

2 The /Data Base Commands

```
                    XYZ Inc Annual Sales

          Name                Sls   Sales (units)        Ten
    Rec  Last        First    Off   YTD   Prev  Pos     Rnk
      8  Fuller      Kelley   W     102    88   O         7
      9  Manning     Kimberly W     102    88   I         8
      7  Matlock     Debbie   W     166   111   O         6
      6  Reyes       Christie W      52     0   I        20
      5  Sanchez     Abraham  W     312   270   Sv        5
     10  Tran        Kim      W      34    10   I        91
     14  Armijo      Christina M    234   112   Sv       13
     13  Atilano     Markl    M     189    47   O        12
     15  Brake       Brian    M      23    25   I        14
     16  Burrows     Holly    M      19     0   I        15
     12  DeStuben    Thomas   M      77    77   I        11
     11  Dyer        Kenny    M     198   121   O        10
      3  Dhanoa      Harpreet E     163   150   I         3
      1  Nokodym     Paul     E     217   205   M         1
      2  Sager       Justin   E     215   175   Sv        2
      4  Sathoff     Jamey    E     199   214   O         4
```

Figure 2.11 Data Base Sorted by Sales Office/Last Name

The commands required to execute this find are as follows:

```
/dqia4.h28<return>
cb53.b54<return>f
```

The /dq (Data Query) is followed by setting the input range (ia4.h28). The second line sets the criterion range and executes the find. Note that the input range included the column headings (second line of the headings only). Figure 2.12 shows the spreadsheet after the execution of this find command. Press the down arrow key to jump from record to record, identifying the Midwest office sales staff. To exit from Query Find, enter the following:

```
<return>q
```

Change the find criterion by entering E or W into cell C54, and press the F7 key to execute a find for another office.

4. Adding an output range requires only that the headings be entered in a suitable area of the spreadsheet and that they be declared as an output range.

```
/ca3.h4<return>a56<return>
/dqoa57.h57<return>
eq
```

The first line copies the column heading to an area at the bottom of the current spreadsheet. The second line executes a /Data Query output and sets the range A57-H57. Only the lower line is declared as a range because it is the range specified previously when the Query Input command was executed. The order of the columns can be changed at this time by rearranging the headings. The final line executes the extract and quits Data Query. At this time, all of the sales records for the East office will be entered below the heading in cells A58-H63.

Using /rea58.h63<return> (Range Erase), the table can be cleared. If cell C54 is changed to M or W, the corresponding records can be extracted by the following command:

/dqeq   Data Query Extract Quit

2 The /Data Base Commands

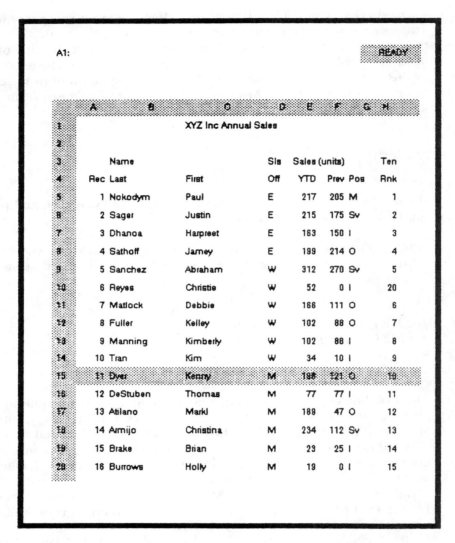

Figure 2.12 Results of Data Query Find

5. After the sort operation, the following steps should be executed to create the file XYZEXTR:

```
/fxvxyzextr<return>a1.h28<return>
/frxyzextr<return>
```

The /Data Base Commands **2**

These are the File Extract values (/fxv) followed by a File Retrieve (/fr) to load the extracted file.

Figure 2.13 shows the summary area of the spreadsheet after these operations have been completed. Data should be extracted in this manner prior to setting up complicated calculations. If the original data base is used, the values in the calculations change when data are sorted.

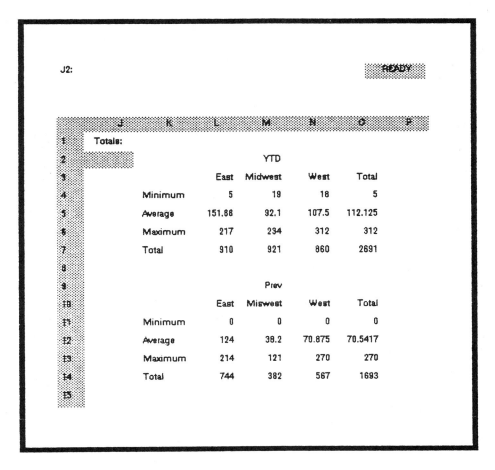

Figure 2.13 Summary Area in Extracted File

53

CHAPTER 3

MACROS

3 Macros

Macros, more than any other Lotus-related topic, have been given an unduly harsh reception. If you work through this book, you will discover two points about macros. First, a macro is *not* a mystical rite reserved for gods and gurus. Second, with a little practice and patience, and the help of this text, you will develop some guru-like abilities through the use of macros.

By definition, a **macro** is an instruction or set of instructions that can be executed with a single command (i.e., by pressing a single key or combination of keys). These key-reducing sequences were initially used in assemblers and other sophisticated software, which may explain why they are now considered difficult. Functionally, a macro provides a way to store Lotus commands and actions in a named range of a spreadsheet. These actions are then executed with a keystroke command, which gives you the ability to write simple programs to perform routine or repetitive tasks.

Lotus Development Corporation, in their 1-2-3 documentation, calls macro operation the "typing alternative." While this statement is true, it understates the power of macros. Anyone who has had to set up a table by manually adjusting the width of twenty columns would view macro ability as more than a typing saver. To accomplish such a task, a macro can be built in a matter of seconds, then deleted after it is run.

This chapter will cover when to use a macro; how to create, save, and execute it; how to debug it with the Step function; how to input into a macro; and how to use the /X commands to round out macros.

WHEN TO USE A MACRO

The question of when to use a macro is not as simple as it might seem. Many users feel they don't need to write macros because their spreadsheets "aren't that complex." Unfortunately, their reasoning is backwards. Very complex or single-function spreadsheets benefit less from macro applications than do simple tasks. The commands that you repeat frequently and the tasks you perform routinely are the ones you want to automate.

One useful macro sets a group of column widths. To set 20 columns to width five, you would enter the following commands. The first command is entered as a label and then copied to the 19 cells below it. The top cell is given a range name so that 1-2-3 knows where to begin.

Command	Comment
'/WCS5~{right}	Sets Width of the current column to 5 and moves cursor to the right.
/c\<return>,\<PgDn>\<return>	Copies the command over 20 cells.
/rnc\a\<return>	Names the range.

When you press the <Alt/A> key combination, 20 columns, starting with the current column, will be reset to width 5. In this case, the macro truly is a typing alternative and also saves time. This process is a classic case of automating a repetitive task.

Conversely, if you plan to do your income taxes with Lotus 1-2-3, there is no need to set up complicated macro functions. The setup would take longer than the time saved. Time should be your deciding factor. If using a macro doesn't save you time or increase your accuracy, then it hasn't provided any advantage.

CREATING AND SAVING A MACRO

The first step in generating a macro is to perform the operation manually and record the steps. This step is the most critical. The commands that follow will generate the table shown in Figure 3.1 in the next section.

Action	Lotus Response
1. Press \<Home>	Cursor moves to A1.
2. Press 2 \<return>	2 entered in A1.
3. Press \<down arrow>	Cursor moves to A2.
4. Press 4 \<return>	4 entered in A2.
5. Press \<down arrow>	Cursor moves to A3.
6. Type @SUM(A1.A2) \<return>	Sum is entered in A3.
7. Press \<right arrow>	Cursor moves to B3.
8. Type TOTAL \<return>	Label entered in B3.

3 Macros

Here, the keystrokes required to accomplish the essential task have been recorded. Once this recording process is completed, the macro has been written. Again, this macro is a list of standard Lotus commands and actions that you use every day. It does not necessarily include any commands you only rarely use; learning those commands is a separate exercise. The commands must be translated into Lotus macro statements and stored for execution. The following listing is the form for the macro previously described:

```
'{HOME}
'2~{DOWN}
'4~{DOWN}
'@SUM(A1.A2)~
'{RIGHT}TOTAL~
```

The only new symbols used here are the tilde (~), {home}, and {down} symbols. The tilde tells the macro that if you were executing the statements, you would press the Return key. The macro executes the Return key command. Cursor movement keys are executed by entering their names in curly brackets ({}). The whole listing can be entered on a single line, but spacing to allow a single operation or group of operations per line makes follow up easier.

Translating the macro into coded format does require learning these few symbols, but this example should show how closely the symbols are related to the actual key sequences. Users sometimes resist advanced techniques, and macros seem to be among of those they are reluctant to try. If you have never tried writing a macro, take a few minutes and enter the example in the next section. Hopefully, you will realize that creating macros isn't that difficult.

All that remains to make this macro executable is to name the starting cell with the /(R)ange (N)ame (C)reate command. There are 27 valid macro names; any single letter can be used as a name in addition to the automatic start macro \0 (zero). The \0 macro will run any time the spreadsheet is loaded (just as an AUTOEXEC file runs when MS-DOS is booted) but cannot be executed manually. If manual execution is also required, then you need to give the starting range two names. The AUTO execution feature will be used later in this book to automatically execute a menu when the Menu spreadsheet is loaded.

Permanent macros can be saved by saving the spreadsheet, which is stored on disk as a range of labels and reloaded with the /(F)ile (R)etrieve command. Even the range name is saved when this procedure is executed.

CREATING A SIMPLE MACRO

In this section, you will generate and execute a simple macro. This process consists of three parts:

- entering all commands as comments into the spreadsheets
- naming the range
- executing the commands

In macro development, the tilde character (˜) is used to automatically execute a Return key command anywhere such a response is required by a Lotus 1-2-3 command. The first sample macro will simplify this discussion.

For this example, erase the spreadsheet:

/WEY   Cursor at A1

Enter the following:

/RNC\A   Cell A1 is named range \A

3 Macros

Then, enter the following commands into the spreadsheet:

Cell	Contents
A1	'{Home}
A2	'/WIR{Down}{Down}{Down}~
A3	'/WIC{Right}{Right}{Right}~
A4	'{Home}
A5	I did it {Down}
A6	I did it {Down}
A7	I did it {Down}
A8	I did it {Down}
A9	It was easy!{Left}{Left}

Execution is easy. Hold down the <Alt> key and simultaneously press the "a" key. It doesn't matter whether or not Caps Lock is on because either upper- or lowercase will start the macro. Figure 3.1 shows how the screen should appear immediately after execution.

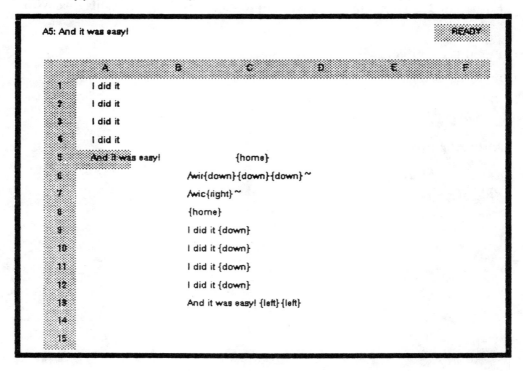

Figure 3.1 After Execution of First Macro

60

A line-by-line examination of the macro's execution should make this process easier to understand. Naming the beginning cell \A defines the starting point for macro execution as well as the command used to start execution.

Macro execution begins in cell A1. Note that an apostrophe (') is shown as the first character in cell A1. The apostrophe is automatically entered if the first character is a label. The {Home} command causes the cursor to move to cell A1 when the macro runs. Execution of the macro continues in cell A2 and moves down until a blank cell is encountered.

The sequence of commands in cell A2 inserts four rows above the macro in the current spreadsheet. The tilde (~) executes the <return> needed to end the Range command. Similarly, cell A3 inserts two columns to the left of the macro. Even though the macro has been moved down and to the right by these actions, it still continues to execute from the top downward. Each cell can contain up to 255 characters with no limit on the number of commands it can contain. For the sake of readability, the macro should be divided the way it was in this example, allowing one action per cell (e.g., insert four rows; insert two columns). A4 returns the cursor to the HOME position, cell A1.

Cells A5-A9 print text in column A, rows one through five. The macro finishes with two {left} commands. Since these commands are executed from column A, they are illegal commands and will cause the infamous BEEP tone to occur. This is a convenient way to let you know execution is complete. (Some macros can take several minutes to execute, so this feature should not be ignored.)

Second Macro Example

In this section, you will enter one more macro from the beginning. In this exercise, you want a macro that will do the following:

- re-enter the data table in Figure 1.2
- set column B to width 5
- set the print ranges and print the table and macro

3 Macros

To define the required macro, the procedure should be "dry run" and the steps recorded. The necessary steps to perform the desired task are as follows:

1. Home the cursor.
2. Insert nine rows to move the macro.
3. Write the labels in column A.
4. Move the cursor to column B (top).
5. Set column width to 5.
6. Enter the numbers and formula in B6.
7. Home the cursor.
8. Enter the print range.
9. Align the printer, Go, Quit print mode.

The beginning point in the macro, cell A1, must be named, for example, as follows:

/RNC\G <return>

When < Alt/G > is pressed, the macro is executed as expected. The table is created in the upper left-hand corner of the spreadsheet and then printed. At this point, special print codes could have been used to change the text style or size; a graph could have been created; another set of data could have been merged from a data file; or any other operation could have been executed along with the macro. Extensive use of these techniques will be made in later chapters.

The table from Figure 1.2 has been re-created in the upper left-hand corner of the spreadsheet in Figure 3.2, which reflects the table and the macro instructions after they have been sent to the printer. When looking at the macro listing, remember that although the apostrophe does not show in the printout or on the screen, it is actually there as the first character in each line because the lines are text.

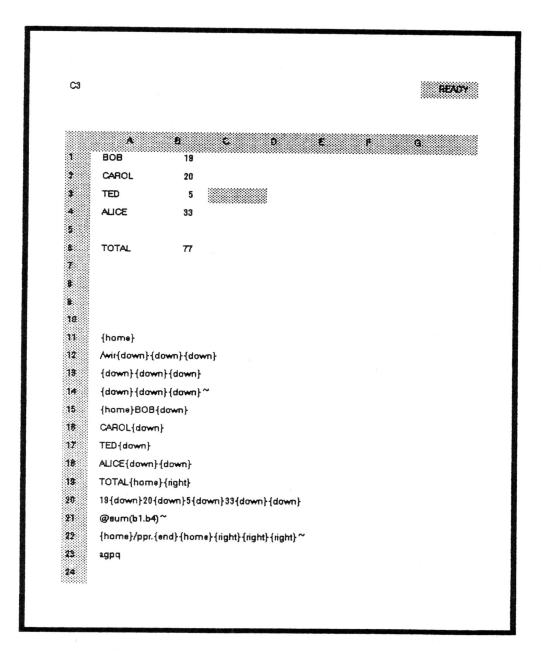

Figure 3.2 Spreadsheet after Running Macro Two

DEBUGGING A MACRO

When you are entering macros, it is best to input them in sections that can be executed and tested individually. Even when breaking macros down to the simpler level, you will find there are still times when you will make a mistake. These mistakes can be hard to find, so a debug mode has been provided. Press <Alt/F1> (the help key) to invoke this mode. When the macro is executed, you will need to press the Space Bar to keep the execution process moving. Step mode is canceled by pressing <Alt/F1> again (a toggle command).

Each time the Space Bar is pressed, another keystroke is entered from the macro range. The commands entered into the command line will be spelled out letter by letter, and all menus will show up in the command line as they are executed. This process can be confusing at first, but with a little practice, it will work for you. Try executing the current macro under the Step mode now so that you will be familiar with the process when you need it.

INPUT TO A MACRO

As data sequences are entered and run in macros, pieces of information are frequently required. These pieces of information can be range names, values, or file and graph information. If you want this information entered at the time of macro execution, rather than in the file, the input command allows that option.

If a question mark (?) is encountered in a macro, Lotus stops and waits for keyboard input before it continues execution. Two additional commands, /XL and /XN, also provide means to input information from the keyboard. These commands will be explained in the next section.

THE /X COMMANDS

A set of true programming commands is provided to round out macro operations. These commands allow you to generate macros with menus and branching, as well as directly input numbers and labels to cells on the worksheet.

Macros 3

Following are the available commands listed with their respective functions:

Command	Function
/XIcondition~	If-Then
/XGlocation~	GoTo
/XClocation	Go To subroutine
/XR	Return from subroutine
/XQ	Quit the macro
/XMlocation~	User-defined menu
/XLmessage~location~	See text
/XNmessage~location~	See text

The /XI command continues execution of the current command if a built-in logical statement is true. If the statement is false, execution drops to the next cell. The end of the current cell can be a branch instruction allowing a conditional test to separate two paths in the program. For example, note the following:

Cell	Command	Action
I100	/XIA1>5~/XQ	Halts execution of the macro if contents of A1 exceed 5.
I100	/XGA100~	Continues execution of the macro at cell A100 instead of I101.
I100	/XCA100~	Continues execution of the macro at cell A100. When an /XR is encountered, return execution to I101.

The /XG command causes the execution of the macro to go to a subprogram or different area of the main program to continue operation. /XC performs the same function but preserves the present location in a buffer. If a /XR return command is encountered, control returns to the cell below the present cell where control was located before the branch.

The /XQ command quits the macro.

65

3 Macros

Menus very similar to the normal Lotus menu structure can be developed with the /XM command. Refer to Appendix D for this information. The /XL and /XN commands print a label above the spreadsheet, prior to requesting a keyboard input. After the input is made, Lotus stores the inputted data in a cell whose location is specified in the location range of the command.

Cell	Command	Action
I100	/XLENTERNAME~a1~	Prints "ENTER NAME" on the command line at the top of the screen and waits for keyboard input. After <return>, text or numbers that were entered are stored in A1 as labels.
I100	/XNENTERAGE~a1~	Same as label input, but contents must be a number.

SUMMARY

- A **macro** is a set of standard instructions that can be automated to a single keystroke.

- The most difficult part of writing a macro is deciding what functions and keystrokes are required. This process is the same as setting up a nonautomated spreadsheet.

- A 1-2-3 macro is a set of commands entered into a named range as labels. It is executed by entering the range name. Valid macro names are single letters or \0 (zero).

- Virtually anything that can be done in Lotus 1-2-3 can be done with a macro.

- The /X commands provide the power to program, which includes keyboard input, looping, and branching.

EXERCISES

1. Write a macro to create a data range in the upper corner of a spreadsheet and fill it with the integers from 0 to 99. The columns should first be set to width 5, and ten rows should be provided in the data range.

2. Expand the macro in Exercise #1 to print the range generated.

3. Modify the macro in Exercise #2 to print a heading on the page; then print the rows of the table in reverse order. The heading should be found in cell K1, and it should read as follows:

   Exercise 3.1 print from macro

3 Macros

4. Modify Exercise #1 again to do the following: sum each row in the data table, convert the equations to actual values, erase the table, move the ten sums to column a, and erase everything in the spreadsheet except the column of values. **Hint**: Use a loop to move down the column of sums, and convert them to numbers with {edit}{calc}.

Answers

1. The following macro starts in cell A12, with a label in cell A11. The macro must be named so it can be executed; the command to name it as \A is:

    /rnc\a<return><return>

    Once the macro has been named \a, it will execute when <Alt/A> is pressed.

    Cell A12 moves the cursor to cell A1. The next five lines each set two columns to width five. The macro is set up this way to allow one line to be entered and copied to the next four lines to reduce typing. Cell A18 executes the data fill. The range a1.j10 sets up the ten-row range as required, followed by the start, step, and stop quantities.

    ```
    A11:  'Macro1
    A12:  '{home}
    A13:  '/wcs5~{right}/wcs5~{right}
    A14:  '/wcs5~{right}/wcs5~{right}
    A15:  '/wcs5~{right}/wcs5~{right}
    A16:  '/wcs5~{right}/wcs5~{right}
    A17:  '/wcs5~{right}/wcs5~{right}
    A18:  '/dfa1.j10~0~1~99~
    ```

2. The previous macro can be expanded so that the entire range can be printed by adding the following:

    ```
    A19:  '/ppra1.j10~gq
    ```

Execute the macro again by entering <Alt/A>.

3. When the required data label is entered in cell K1, as in the following, it is a simple matter to print the data label as a heading. Cell A20 accomplishes this task. Cells A21-A30 print the rows of the table, and the (q)uit in A30 exits print mode.

```
A19:  '{home}
A20:  '/pprk1~g
A21:  'ra10.j10~g
A22:  'ra9.j9~g
A23:  'ra8.j8~g
A24:  'ra7.j7~g
A25:  'ra6.j6~g
A26:  'ra5.j5~g
A27:  'ra4.j4~g
A28:  'ra3.j3~g
A29:  'ra2.j2~g
A30:  'ra1.j1~gq

K1:  'Exercise 3.1 print from macro
```

4. The result of running the macro described in the problem statement is a column of integers, starting with 450 in A1 and ending with 540 in A10, each 10 larger than the one above it. The macro that follows accomplishes this task.

Cell A20 turns off the recalculation mode so the loop counter increments only when needed. A21 sets cell I1 to 10 (the ending count) and I2 to I2+1, which makes the initial value of I2 equal to 1. Every time a {calc} command is executed, this count will go up by one.

A22 sends the cursor to cell K1 and sums the row. A23 copies this summation to the cell below itself, executes the {edit}{calc} sequence to convert the current contents [@sum(a1.k1)] to the value 450, moves the cursor down one cell, and executes the {calc} to increment I2.

69

3 Macros

A24 compares the contents of cell L2 to L1 to see if all ten rows have been summed. If L2 is less than or equal to L1, there are more rows, so the /xg (goto) portion of the statement is executed. This result causes the macro to branch back to A23 and sum the current row.

Cell A25 erases the table and moves the column of numbers to the home position. A26 erases the column with the loop variables, and A27 erases the rows below the data column. Note that this process includes removal of the macro itself.

```
A11:  'Macro1
A12:  '{home}
A13:  '/wcs5~{right}/wcs5~{right}
A14:  '/wcs5~{right}/wcs5~{right}
A15:  '/wcs5~{right}/wcs5~{right}
A16:  '/wcs5~{right}/wcs5~{right}
A17:  '/wcs5~{right}/wcs5~{right}
A18:  '/dfa1.j10~0~1~99~
A19:  '{home}
A20:  '/wgrm
A21:  '{goto}l1~10~{down}+l2+1~
A22:  '{goto}k1~@sum(a1.j1)~
A23:  '/c~{down}~{edit}{calc}{down}{calc}
A24:  '/xil2=l1~/xga23~
A25:  '/rea1.j10~/mk1.k10~a1~
A26:  '/wdc.{right}{right}~
A27:  '/wdr.{pgdn}{pgdn}~
```

CHAPTER 4

FILE OPERATIONS

4 File Operations

To understand how data are stored and recalled using Lotus 1-2-3, it is necessary to go beyond the simple /FS (File Save) command. In this chapter, you will **import** and **extract** data, so that 1-2-3 can be used in conjunction with other software, and merge files with the /FC (File Combine) utility.

Saving data to disk should not only be considered when that data needs to be archived. Even if a permanent disk file is not needed, data should be saved to a file every ten minutes as a buffer against lost time. PCs are very reliable, but power failures and accidental erasures do occur. If data are saved to disk every ten minutes, you won't need to spend hours reconstructing the information if a problem occurs.

There are three ways to write data to a disk: using the File Save, eXtract, or Print File command. While similar in nature, these commands have different uses. **File Save** stores whole spreadsheets for later retrieval. **eXtract** stores only partial spreadsheets for retrieval or combining with other spreadsheets. **Print File** creates an ASCII file that can then be read by other applications. When File Save is used, all spreadsheet and graph settings are saved with the file.

Spreadsheet input includes retrieving data with the **Retrieve** command, merging extracts with the **Combine** command, and importing data from ASCII files with the **Import** command.

GETTING DATA TO A DISK

The simplest of all the disk commands is the /FS (File Save) command. On the first execution of this command, you are prompted to enter a data file name. For subsequent saves (after a file has been retrieved from the disk), the prompt gives you the option of using the "current" name, that is, the last name used in a Save or Retrieve command. If the file name to be used already exists on the disk, you have the option to cancel the command, not modifying the file on disk, or replace the file on disk with the current file.

File Operations 4

A name in 1-2-3 has the same restrictions as a name in DOS but with a few more limitations. Lotus does not support path entry as part of the name; the path is assigned by the Directory subcommand. This information can be viewed with the following command:

/FD   (File Directory; the current directory is displayed)

In most cases, even when a hard disk is used, the directory will be B:\. The backslash (\) delimiter is used to separate subdirectories, which can be used on hard or floppy disks with DOS 2.0 or 3.0. To change the directory at this point, you need to enter a new path, but most people prefer to leave the directory as B:\. One quirk of 1-2-3 shows up here: the directory cannot be changed from B:\ if there is no disk in drive B.

New users frequently confuse *displaying* the current directory with *listing* it. The /FD command is used to identify or change where files are located; the actual listing of files requires the following command:

/FL   (File List; files listed by type)

Extensions to data file names are reserved for use by the program. All Lotus release 1 spreadsheet files have an extension of .WKS; release 2 spreadsheet files have .WK1 extensions; files created by Print File have .PRN extensions; and graphs have .PIC extensions. If you want to import a file, it must have an extension of .PRN. Lotus release 2.0 will read a release 1(A) .WKS file but will change the extension to .WK1 when it is saved. This change allows the program to distinguish between the two files; release 1 files can be used by release 2 but not the reverse.

Note: direct access to subdirectories is possible using the SUBST command in DOS (3.0 or higher) — see Appendix G.

4 File Operations

File Extractions

Figure 4.1 provides a useful example for studying file extraction. A manufacturer of widgets has factories in Albany, Chicago, Dallas, Houston, Ontario, and Seattle. The spreadsheet provides a monthly summary by product line for the IDGET company. The staff size and production rate are used to calculate widget production quantities.

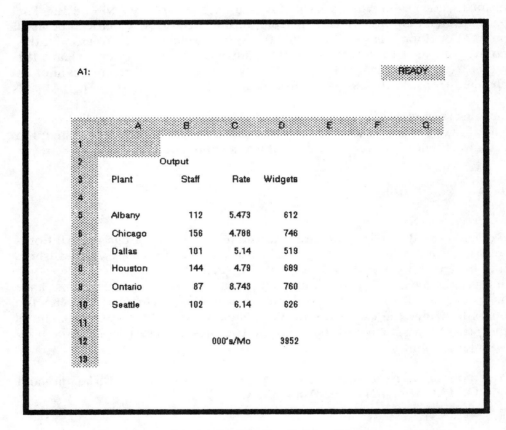

Figure 4.1 Production Output Table: Widgets

File Operations 4

Enter the spreadsheet from the keyboard and save it as file "Examp4_1." The following commands will eXtract the production quantities from this file for use in another file. Be sure the file is saved; you will use it again in this exercise.

User Keystrokes   1-2-3 Response

<F5>D3<return> (goto D3)   Cursor on Widget at D3
/FXF(eXtract Formulas)   Enter extract file name:
Examp4_2<return>   Enter extract range:
<down><down><down><down>
<down><down><down><down>
<down><return>   Range Painted, File saved

Now that you have created a file with the eXtract command, note that it is a spreadsheet and can be retrieved using the /FR command. Figure 4.2 shows the result of this retrieval.

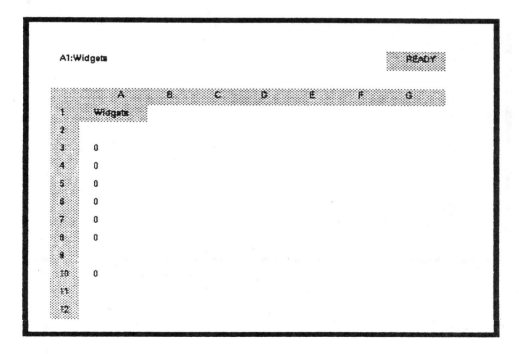

Figure 4.2 Data Contained in Extracted File

4 File Operations

Note that all of the cells that held the production quantities show a value of zero. Recall also that when you executed the eXtract command, you selected formulas. Move the cursor to one of these cells, and the formula will be there. Since the relative cells are empty here, the cell values are zero.

Repeat the steps used earlier to create Figure 4.1, substituting

/FXV   (File eXtract Values)

in the second line. Figure 4.3 shows the updated "examp4__2" after it is retrieved from disk. The actual cell values replace the formulas. This is the most frequently used method for extracting files.

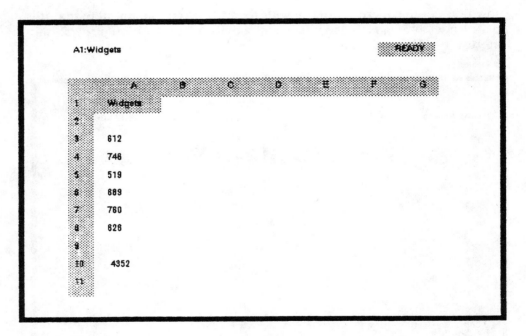

Figure 4.3 Data Contained in Extracted Values File

File Operations 4

File Printing

The last file storage option to be considered is **file printing**. By now, you should have sent a spreadsheet to the printer. After pressing /P, you are offered the following response on the command line:

```
Printer  File
```

Up to now, your choice probably has been Printer. Retrieve "examp4_1" again and print it to an ASCII file, using the second option, File. With the spreadsheet corresponding to Figure 4.1 on your screen, enter the following:

User Keystrokes	1-2-3 Response
/PF (Print File)	Enter Print File Name:
examp4_1<return>	Print options Menu
R<Home> .<End><Home><return>	Set print range
OOUQ (Options Other Unformatted Quit)	Print Options Menu
	Cursor on Options
GQ (Go Quit)	Disk accesses
	File Printed

Note that it is acceptable to use the same file name for both the data file and the print file. Both will be stored on disk with different extensions. Furthermore, all the commands in the Print menu have the same function here as when printing in the standard fashion. Two or more ranges can be sent to the file with a (P)age Break or (L)ine between them. Options can be changed between sections. The file is not closed until the (Q)uit is executed.

Save the file again under the name "examp4_3." Exit Lotus, and at the DOS prompt, change the directory to the data file location. In most cases, this location will be B: Now try the following commands:

User Keystrokes	1-2-3 Response
B:<return>	B:>
type EXAMP4_1.prn	Figure 4.1 prints to monitor
copy EXAMP4_1.prn lpt1:	Figure 4.1 is sent to printer

77

4 File Operations

As you can see, the file outputted by the /PF command is a printable ASCII file made up of standard characters. This file can be passed to word processors, printer utilities, desktop publishers, or any other suitable software package. It can even be read as data into a BASIC program, eliminating the need to write data handlers to generate data files.

RETRIEVING DATA FROM A DISK

In addition to the standard /FR (File Retrieve) command for recalling saved spreadsheets, data can be **(I)mported** or **(C)ombined**. The Combine command provides a means for merging all or part of a spreadsheet into a disk file with the spreadsheet that is currently active. The Import command allows files created in other applications to be converted into spreadsheet format.

File Combine

The **File Combine** command can easily be demonstrated. Erase the current worksheet and merge the previously eXtracted file with the following commands:

User Keystrokes	1-2-3 Response
<F5>D2<return>	Goto cell D2
/FCCE (File Combine Copy Entire)	Enter Name of File to Combine:

Using the cursor keys, move the command cursor to file name "examp4__2" and press <return>. The range eXtracted from "examp4__1" will be copied into the current spreadsheet, starting at cell D2.

The **Copy Entire File** option loads a copy of the saved or eXtracted spreadsheet, with its upper left-hand corner at the current cursor location. The most common error made here is misplacing the cursor, which can allow the merging data to overwrite a necessary part of the spreadsheet.

To illustrate the **Named Range** option, you must have a spreadsheet with a range named. Retrieve the "examp4_1" file again and name the WIDGET output column using the previously described Range Name commands:

```
/FRexamp4_1 <return>
<F5>D3<return>
/RNCWIDGETS
<down> (9 times)
<return>
/FS<return>R (File Save "same name" Replace)
/WEY (erase worksheet)
```

You now can combine the range named WIDGETS into the blank spreadsheet with the same commands you just used to merge a whole file:

```
/FCCN WIDGETS <return>
```

At the prompt, you will see the following message:

> Enter Name of file to combine:

The cursor keys can again be used to highlight the file name "examp4_1." Press <return> and Lotus will open the data file, locate the range named WIDGETS, and insert it into the current spreadsheet at the cursor position. This method is limited to transfer of formulas, so all of the values are zero, as seen previously in the eXtract example.

IMPORTING DATA

Just as text is printed to an ASCII file for output to other software, you can input text into a spreadsheet using the **/File Import** command. This command allows data files generated in high-level language software to be transferred into Lotus. Data transferred from other operating systems (passed across modems, exported from DBMSs, or generated with editors and word processors) also can be imported. Additional information on data base imports can be found in *dBase III PLUS Power Tools* by Rob Krumm (MIS: Press, 1986).

4 File Operations

Data import can be accomplished in two formats: number and text. Since the Import function will ignore any text in a numbers file, text mode must be used if there is any text material. In either mode a <return> is the record delimiter, but in number mode, commas are treated as field delimiters. Data are entered into the cursor cell, in text mode, until a <return> is encountered. Then control is transferred to the cell below. No field separation is used; everything is compressed into the first field and must be **parsed** (separated) later. In numbers mode, commas or semi-colons separate fields that are placed in separate columns.

To demonstrate this function, use the Copy command to create an ASCII file for import. The same file could be created using a word processor, EDLIN, or any other text processor. Carefully copy the following commands. If you make an mistake, start over. As stated earlier, <return> indicates pressing the Return key, and <Ctrl/Z> indicates holding down the Ctrl key while simultaneously pressing the Z key.

```
copy con: b:examp4_4.prn <return>
1 <return>
2 <return>
3,4,5 <return>
6;7;8 <return>
one <return>
two <return>
three,four,five<return>
<Ctrl/Z> <return>
```

Using the DOS TYPE command, you can verify that the information has been entered as expected:

Type b:examp4_4.prn

File Operations 4

Your screen should display the following:

```
1
2
3,4,5
6;7;8
one
two
three,four,five
```

Now import this data file into Lotus. Load 1-2-3 and execute a File Import Text command:

/FIT <examp4_4> <return>   (type in angle brackets around "examp4_4")

With most systems, the file will load directly. If an error message occurs, press <return> again, and the file should load with the warning "part of file missing." Figure 4.4 shows the result of this load:

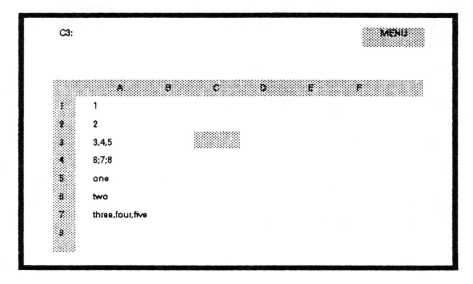

Figure 4.4 File Import Example

4 File Operations

The four records containing multiple fields were all entered in single cells in column A. This result is unfortunate but expected because the delimiters chosen (comma, semi-colon, and slash) are themselves ASCII characters and are undistinguishable from the rest of the elements in the record. The Parse statement described previously allows a record to be divided into separate fields, but only if the field widths are standard (i.e., if each of the eight records here had exactly three fields, if each field were five characters wide, and if each field were padded with spaces to fill the blanks). Most application programs send text files suitably constructed. If custom applications are being developed, then caution must be exercised.

Repeat this action for the same file in numbers mode:

/WEY   (Erase worksheet)

/FIN <examp4_4> <return>   (type in angle brackets around "example4__4")

You can see that all the character entries were ignored, so only four records appear. Both the comma and semi-colon were accepted as delimiters, so the second pair of records was split into three columns.

A simple set of rules can be applied if applications are being developed to pass data from programs in higher level languages to Lotus spreadsheets:

1. Save all data files in uncompressed ASCII mode.

2. Send string (text) and numerical data in separate files.

3. Each string record should be followed by a <return> and will be imported to a line of its own.

4. Data can be imported into a table format. Separate elements in a single row with commas or semi-colons. End the row with a <return>.

File Operations 4

SUMMARY

- File Save, Extract, and Print File all save data to disk for different purposes:

  1. **Save** retains complete spreadsheets.
  2. **Extract** retains excerpts.
  3. **Print File** saves data to ASCII files.

- Spreadsheets can be merged with the **Combine** command.

- The **Directory** command shows where data files are stored.

- The **List** command displays files on the current directory.

- Either numbers or formulas can be **eXtracted**.

- Once a portion of a spreadsheet is extracted, it is itself a spreadsheet.

- Retrieve, Import, and Combine commands allow data to be entered into spreadsheets as follows:

  1. **Retrieve** loads whole spreadsheets.

  2. **Combine** merges whole spreadsheets or extracted information into the current spreadsheet.

  3. **Import** allows data from programs other than 1-2-3 to be entered into the current spreadsheet.

EXERCISES

Case Study Two (Continued)

The exercises that follow make use of the data file created in Case Study Two at the end of Chapter 2. Either recall the data file you created there, or enter the data in Table 2.1 into a spreadsheet per the instructions in the Chapter 2 exercises. That file was stored as "case2"; if your file name is different, substitute it in the exercises that follow.

1. Create individual spreadsheets for each sales office, containing all of the information about the individual office.

4 File Operations

2. Using the **Print File** option, output the entire data file, including information about all three offices, to a file named "case2.prn." Exit Lotus and view this file on the monitor. Send this file to the printer.

3. Import the print file created in Exercise #2 into a new spreadsheet. Explain the differences between this file and the original.

4. Using the **Parse** data base command, restore this data to spreadsheet format. (**Note**: release 2 only.)

Answers

1. Sort the data base using Sls Off as the primary key, and sort in descending order. All of the West office entries will be on top. Use the following commands to store this information in a spreadsheet file named "case2_a" (case study 2 file a):

   /fxvcase2_a\<return>a1.i12\<return>
   /wdra5.a12\<return>

   The second command, Worksheet Delete Row, removed all the West office entries, leaving the Midwest on top. The next commands create a file, "case2_b," with the Midwest data:

   /fxvcase2_b\<return>a1.i14\<return>
   /wdra5.a14\<return>

   This leaves the East Coast office data on top. The commands that follow create "case2_c," the East Coast office data file, and restore the complete data file for the next exercise:

   /fxvcase2_c\<return>a1.a10\<return>
   /frcase2\<return>

2. The following commands print the entire file to a .PRN file with the same name as the original spreadsheet file (case2.prn).

   /pfcase2\<return>
   ra1.i51\<return>
   agq

The first line enters the file name for print; the second line enters the range; and the third line aligns the print counter to top of form, prints, and exits print mode.

The next two parts of this problem require you to exit from Lotus. Change to directory B: or to the directory where data files are located:

`/fdb:<return>`

Verify that the data file is present:

`dir case2.*<return>`

DOS will respond with a directory listing showing only the following:

    case2.prn
    case2.wk1   (.wks for release 1 users)

and the amount of space left on the diskette.

View this file on the screen by entering the following:

`type case2.prn<return>`

The file will appear on the screen exactly as it would have been printed on paper if /Print Printer had been used instead of /Print File.

Send this file to the printer using the following command:

`copy case2.prn lpt1:<return>`

The file will be sent to the printer.

3. Return to 1-2-3 and import the .PRN file with the following command:

    `/fitcase2<return>`

4 File Operations

There are several notable differences:

- The first line of data is in cell A6, not A2 as it was when it was printed. 1-2-3 inserts four rows at the top of the page (default value) when the file is printed. When it is read back, these rows are present.

- The data starts farther to the right than before in each row. Just as the rows were added at the top, four spaces were added to each row by the Left Print option's Margin Left command.

- When the cursor is moved to A10, all of the information for this row is contained in that cell. In data base terms, the entire record is contained in the first field. Cells in columns B-I are empty. The characters 1-2-3 uses to separate columns of data are not recognized in the ASCII format the file takes when it is printed. On import, these characters have been turned into spaces with no column separation inherently provided.

4. Once a data base file is imported, as in the previous problem, it is suitable for parsing. The major requirement is that each row has fields that are consistent in width and format. This requirement will always be needed when files are printed from 1-2-3 (because of its columnar nature) or when they are generated by a DBMS (because of the field/record structure). Two parse commands can be used to restore this file: one for the column headings and one for the data. Some data should be moved in preparation:

```
/fitcase2<return>            (F)ile (I)mport (T)ext
/wdra1.a5<return>            (W)orksheet (D)elete (R)ow
/ma4.a30<return>a48<return>  (M)ove
<F5>a48<return>              GoTo A48
```

File Operations 4

After these commands are executed, the spreadsheet will have been reentered, the extra rows at the top removed, and the lower line of the heading and data moved to a work area below the file. The next step is to parse the heading and return it to row 4:

/dpfc   (D)ata (P)arse (F)ormat-line
                            (C)reate
I.{down}<return>   (I)nput-range
oa4<return>g   (O)utput-range
/rea48.a49<return>   (R)ange (E)rase
<F5>a50<return>   GoTo A50

The format line created by the first command line tells 1-2-3 how much data is in each column and what type of data to expect. The command line can be created manually by an operator or automatically as was done here. This line can also be edited prior to the actual parse. In this case, each region was declared as a label. The command line has a capital L in a position corresponding to the start of each label in the heading line. There is an additional fill character (>) for each letter in the label and an asterisk (*) for each space after the last character and before the next label.

The input range set up in the second command line includes the format line and all subsequent lines to be parsed. The output range tells 1-2-3 where to copy the parsed information. Once the heading is parsed and in place, the data can parsed in the same manner:

/dprfc   (D)ata (P)arse (R)eset
                            (F)ormat-line (C)reate
i.{end}{down}<return>   (I)nput-range
oa4<return>g   (O)utput-range
/re.{end}{down}<return>   (R)ange (E)rase

The format line created this time has a mixture of labels (L>>>) and values (V>>>):

|****************V*L>>>>>*****L>>>>>*****L*****V>****V>*L*****V

87

4 File Operations

Since there is the same number of groups, the parsed data will fill the columns below the heading labels parsed in the previous example.

Since the first line was not parsed, it no longer matches up; it must be replaced. The following cell entries will replace that line:

```
B3:  'Name
D3:  'Sls
E3:  'Sales (units)
H3:  'Ten
```

All that remains is to adjust column widths (4, 12, 12, 4, 9, 9, 3, and 5 were used here) and range label adjustments (right- or left-aligned column headings) if desired. The top-of-page heading and the legend will remain in a slightly different format from the original file, but since they are only text, this difference does not matter.

CHAPTER 5

CREATING AND SAVING GRAPHS

5 Creating and Saving Graphs

Most Lotus users quickly find a need to graph results, and graphing capabilities are among the most exciting benefits of the Lotus package. Data relationships can be displayed by pressing a few keys. Instead of tedious hand-graphing and scaling, in seconds the analyst can make accurate plots of up to six variables. Once data are entered, analysts can use the Graph key to instantly replot new data.

If you have never created a graph, you may want to review this section in a more elementary text or in the Lotus manual. Figure 5.1 provides an overall view of the **Graph** command structure in Lotus 1-2-3. It bears a strong resemblance to a tree, with the **Graph** command (/G) at the trunk. This chapter covers the boldfaced portions of the figure. Later in this book, you will need to reset all or part of a graph within a macro, fully annotate graphs using **Options**, and switch active graphs using **Name**.

5 Creating and Saving Graphs

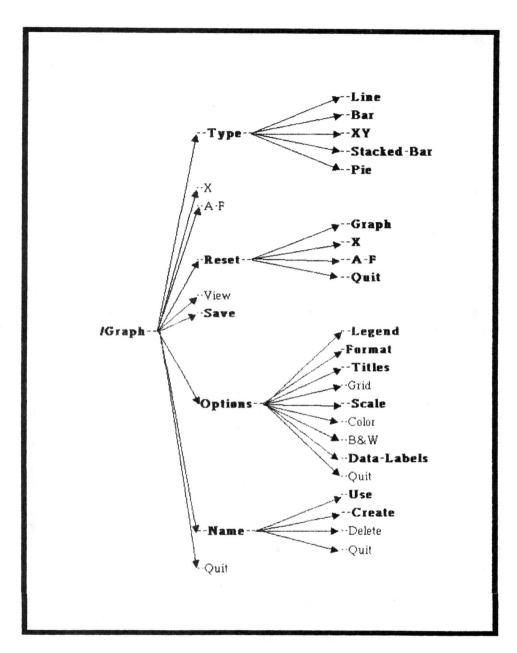

Figure 5.1 Lotus Release 2.01 Graph Menu Tree

5 Creating and Saving Graphs

GRAPH TYPES

The initial Graph menu appears after you enter **/G** . Available options include the following:

- Type
- X
- A-F
- Reset
- View
- Save
- Options
- Name
- Quit

The following sections provide information on Graph commands that will be used later in this book. The Type submenu allows selection of either **line**, **bar**, **x-y**, **stacked bar**, or **pie** graphs (see Figure 5.2).

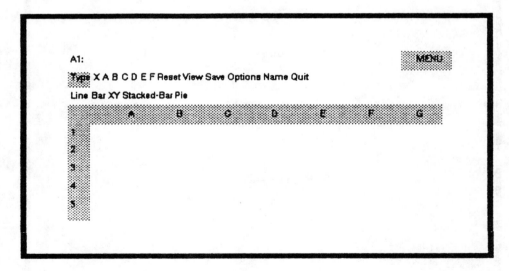

Figure 5.2 Lotus Printgraph Menu

Creating and Saving Graphs 5

Line Graphs

Line graph mode provides two-dimensional output with the following criteria (default settings):

- X-axis points are equally spaced, one per cell in the range. Points are plotted in the order in which they appear in the spreadsheet.

- Only contiguous points in a range are connected when a line on the graph is specified. Blanks in the data range cause discontinuities in the graph.

- The X-axis on the graph starts with the first cell in the range, ends with last cell, and is in the same order as the range.

- One line is added for each Y-axis range (up to 6).

- The Y-axis starts with the lowest point in any range and ends with the highest point.

- Points between the lowest and highest points are proportional.

Since a picture is worth 10K words, a visual example should help you more fully understand graphing. The data table in Table 5.1 will be used throughout this chapter to explore the different features of graphing. The data represent supply and demand projections based on number of units.

Units	Supply	Demand
1	1	20
2	3	19
5	5	18
10	7	17
20	9	15
50	11	14
100	13	12
200	15	10
500	17	9
1000	19	7

Table 5.1 Supply vs. Demand Data

5 Creating and Saving Graphs

Enter the data provided into a spreadsheet, starting in cell A2, and use the following directions to create a line graph showing the intersection of the supply/demand curve. Remember, to paint a range, you must first place the cursor in one corner, and enter a period (.); then use the arrow keys to move to the opposite corner, and press <return>.

1. Type /G (enter graph mode)
2. Type X, paint range A2-A11, press <return>
3. Type A, paint range B2-B11, press <return>
4. Type B, paint range C2-C11, press <return>
5. Type V (view)

The line graph in Figure 5.3 represents what the screen should display. If no picture appears on the screen, and no one has run a 1-2-3 graph on the system before, then check the installation guide to ensure proper drivers are in place. Hercules-type graphics cards must be set to Full mode prior to running graphics, which can be accomplished with software provided with the card.

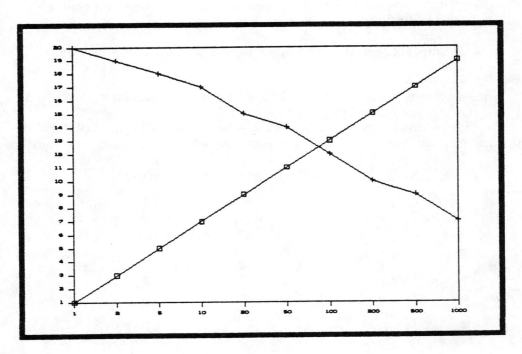

Figure 5.3 **Supply and Demand Line Graph**

Creating and Saving Graphs 5

A closer look at Figure 5.3 provides insight into how 1-2-3 processes graphic output. There are ten labeled points on the X-axis—the contents of the ten cells in the range A2..A11. The only limit to the number of points Lotus will plot is the amount of available RAM.

Two sets of points are plotted on the Y-axis: the A and the B ranges. Scale for the Y-axis will be covered later in this chapter. The points on the X-axis, however, are evenly distributed over the plot range on the screen. Points are plotted in the order in which they appear and not to scale along the axis. In essence, X values are only labels in a line graph.

Bar Graphs

Now that the data are entered and the graph ranges are set, it is simple to examine the different presentation types. To change the present graph to a **bar** graph, type the following:

```
<return>           (returns to the Graph menu)
/TBV               (Type Select, Bar, View)
```

Figure 5.4 shows the same data set in histogram (bar graph) format. Each point on the X-axis has two bars associated with it (later, a legend will be added to distinguish them). Neither axis has changed.

5 Creating and Saving Graphs

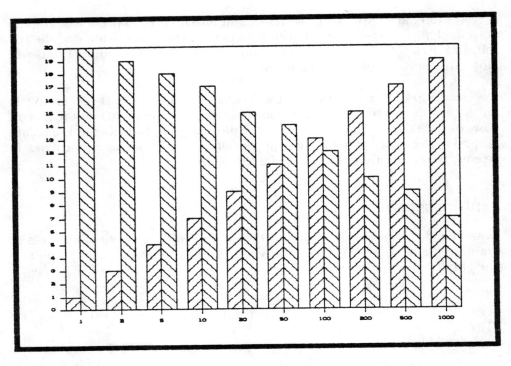

Figure 5.4 Supply and Demand Bar Graph

Stacked Bar Graphs

In some cases, a **stacked bar** format is preferable. Figure 5.5 is the result of entering the following:

```
<return>
/TSV          (Type Select, Stacked Bar, View)
```

If, instead of supply and demand, the columns represented sales by salespersons 1 and 2, the stacked bar would represent total sales. The individual bars are divided into two segments representing the sales for each individual. As before, the X and Y values have not changed.

Creating and Saving Graphs 5

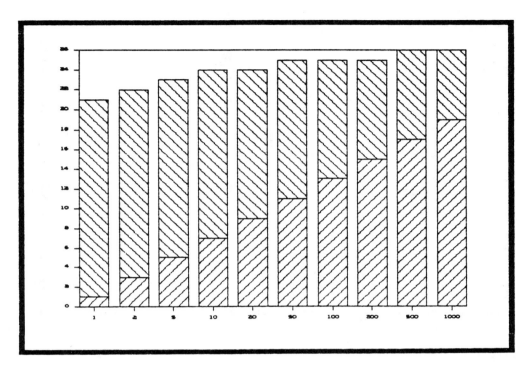

Figure 5.5 Stacked Bar Graph

Pie Graphs

The **pie** graph is significantly different from all the other graphs. For each value in the X range, one slice of the pie is labeled with the X value followed by the A range (B2.B11) value in parentheses. A pie graph, then, is only valid for data sets with paired values. Figure 5.6 is representative of the screen display resulting from the following commands:

```
<return>
/TPV          (Type Select, Pie, View)
```

5 Creating and Saving Graphs

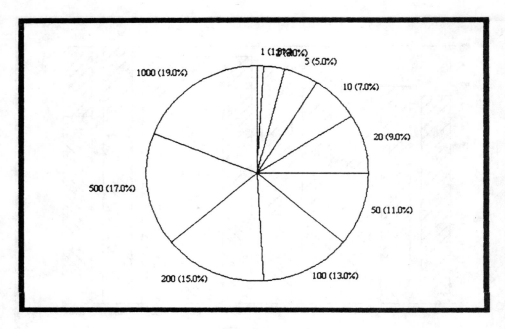

Figure 5.6 Pie Chart Example

X-Y Graphs

The **X-Y** graph is one of two areas in 1-2-3 graphing that causes the most confusion. The following commands will turn the present data set into an X-Y graph (see Figure 5.7).

```
<return>
/TXV              (Type Select, X-Y, View)
```

Creating and Saving Graphs 5

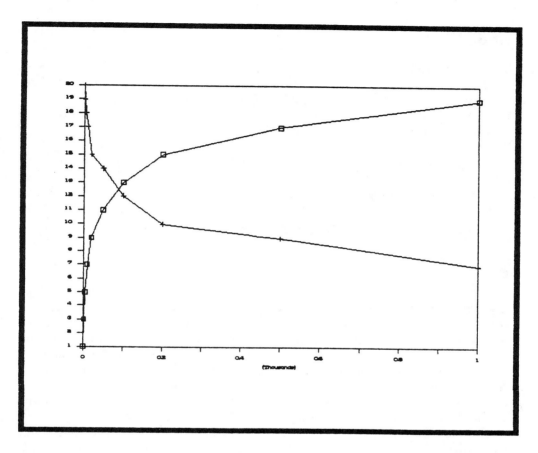

Figure 5.7 Supply and Demand X-Y Graph

Clearly, a significant change occurred in the X-axis, but the Y-axis has stayed the same. Not only has its scale changed to units of 1000, but its spacing is now proportional. The line graph makes both curves look linear, but the X-Y graph reveals their true exponential nature. For most mathematical applications, the X-Y graph is the best choice, although the default setting is Line graph mode.

5 Creating and Saving Graphs

RESET

The **Reset** function is often overlooked. Any one or all of the data ranges (A-F) can be removed from the present graph with the Reset command. Frequently, beginners try to remove a line from a graph by erasing the range and find that the legend will not go away. The Reset command is needed to completely eliminate one of the Y ranges.

The Reset submenu allows you to eliminate all graph settings or one of the axis ranges. This feature is useful in macro programming because Lotus retains the last graph range setting, and there is no way for a macro to verify the current location of that setting. The macro can reset all or part of the graph, and then enter the desired settings.

SAVE

The **Save** command is used to create a printable data file for use by the Print function. This process is the second and most significant area of confusion for Lotus users. After data are entered and a graph is created, the Save command creates a file with a .PIC extension. Remember, spreadsheets have extensions of .WKS or .WK1, depending on the version used. Graphs, as pictures, have .PIC extensions. ASCII files created by the Print File command have .PRN extensions. For convenience, the same file names (with different extensions) can be used for all three types of files, and Lotus will keep track of them for you. (For details on extensions, consult a DOS manual.) The graph can be printed with the Print Graph function, but *it is not yet a part of the spreadsheet.* If you do not save the spreadsheet (using **/File Save**), you will not save the graph as part of the spreadheet.

There is no way to recall a .PIC file and recover the lost graph if the File Saving procedure is ignored. For best results, follow this simple rule: *after every /GS (Graph Save), perform a /FS (File Save).*

OPTIONS

Several menus fall under the **Options** heading. **Legend** allows multiple curves to be identified with a short comment at the bottom of the graph. **Format** lets you change the way the curve is presented. Through the **Titles** menu, two title lines can be added to the top of the graph, and the axis can be labelled. **Scale** allows changes in the graph's data distribution. **Data-Labels** allows a symbol to be substituted for the standard graphic on each graph line, and the symbol can include a label.

Legend

The **Legend** option allows a label for each selected range to be placed under the graph. Since only one line is available for legends, it must be divided among all the ranges. Unfortunately, as more ranges are selected, labels must be shorter. To continue the demonstration, enter the following:

\<return\>	(returns the screen to the Graph menu)
TL	(Type, Line returns to line graph)
OLASUPPLY\<return\>	(Options, Legend, A range, Supply)
LBDemand\<return\>	(Legend, B range, Demand)
QV	(Quit, View)

The supply and demand graph is again visible, with legends for the two lines displayed at the bottom.

Format

The **Format** option allows modification of visual symbols used to generate the graph. Each curve on the graph contains both lines and symbols. The format command allows each curve to be represented by lines, symbols, both, or neither.

5 Creating and Saving Graphs

Choosing **Line** causes X-Y and Line graphs to be represented by a line only. All contiguous points are connected. Blank cells in the graph range cause breaks in the line. The **Symbols** option causes each point in the range to be represented in the graph by the particular symbol for the plotted range. The **Both** option superimposes these two formats. The **Neither** option plots a graph with no visible points. The **Data-Labels** command allows other symbols to be substituted where the original symbol would have appeared.

Try these various options by executing the following:

OFALQQV   (O)ption (F)ormat (A)range (L)ine (Q)uit (Q)uit (V)iew

Substitute (B)oth, (S)ymbols, and (N)either for (L)ine.

Titles

The Titles option allows lines of text to be displayed as graph or axis labels. The **First** and **Second** commands in the Titles menu allow entry of two lines at the top of the graph. **X-axis** and **Y-axis** are labels describing the variables and units of the two Cartesian scales. The following sequence labels both the axis and the graph:

```
<return>
OTFSupplyDemandCurve<return>           (Titles, First line)
TSExample2<return>                     (Titles, Second line)
TXUnits<return>                        (Titles, X-axis)
TYSupply/Demand values<return>         (Titles, Y-axis)
QV                                     (Quit, View)
```

It is not necessary to enter a label directly into the menu. Entering a backslash (\) followed by a cell address will cause the contents of a cell to be used as a label. Not only does this strategy allow automatic graphing through macros, but longer labels can be entered in this manner.

Grid

Grid lines may be added to the graph with the **Grid** command. Horizontal lines, vertical lines, or both, can be added to the graph as required. Figure 5.8 shows the current graph after the addition of Grids (both types of lines) and Legends:

```
<return>
OGBV                    (Options, Grid, Both, View)
```

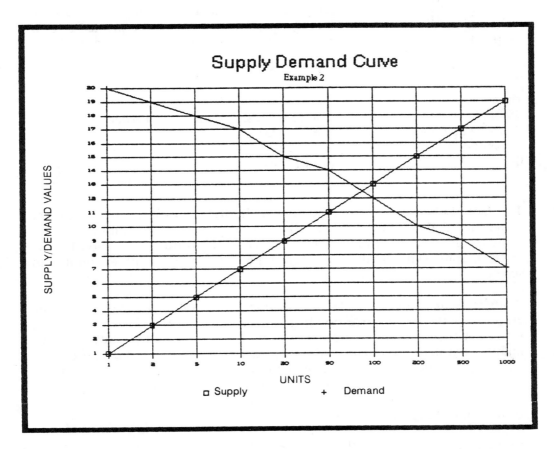

Figure 5.8 Grid Line Example

5 Creating and Saving Graphs

Scale

Manual selection of the X-axis and Y-axis starting points and range can be made with the **Scale** command. The default setting allows Lotus to automatically select ranges based on the maximum and minimum values in the graphs to be plotted. This feature is advantageous for many reasons. In a data set with small variations, the minimum and maximum values will be represented with full-scale deflection and will appear to vary greatly. Manual selection of scales keeps the values in perspective. If two or more graphs are to be compared, they may be easier to read if the scales are the same.

If a large number of points are graphed, the data labels in X overlap. The **Skip** option allows only a predefined part of the labels to be printed (e.g., a Skip value of 3 prints every third label). The following commands cause the Y scale to run from 0 to 50:

```
<return>
OSYMLO              (Options, Scale, Y, Manual, Lower, 0)
U2550<return>       (Upper, 25)
S3 <return>         (Skip 3)
QV                  (Quit, View).
```

The **Data-Labels** option allows columnar information from the spreadsheet to be included on the graph. A row of symbols can be substituted for the standard graphic by filling a data label range. Substituting a period for the larger graphic symbol allows display of a curve as a series of dots plotted to scale.

A single value can be placed in a data label range, and it will show up on the graph near the point corresponding to its position in the range. The effect is to have the label "float" near the curve at that point. The following commands complete the current graph by adding **floating labels** to the curves. Note that to enter these labels with the following sequence, you must exit graph mode and include them on the spreadsheet:

Enter "SUPPLY" in Cell D4.

Enter "DEMAND" in Cell E4.

Creating and Saving Graphs 5

/GODAD2.D11<return>A   (Options, Data-labels, A range
                              D2-D11, Above)
BE2.E11<return>AQQ   (B range E2-E11, Above, Quit,
                              Quit)
V   (View the graph)

Figure 5.9 is the final graph with the modified scale, floating labels, and the X-axis, which is labeled only at 1, 10, 100, and 1000.

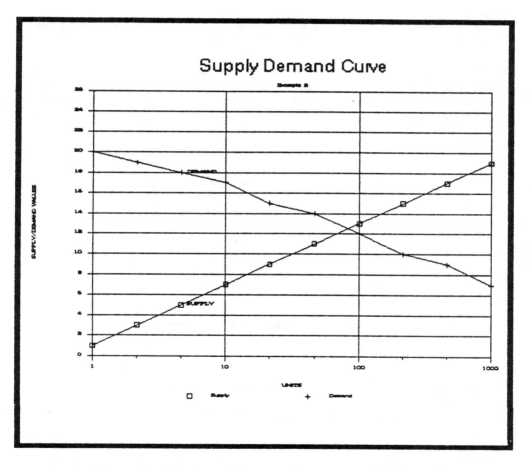

Figure 5.9 Supply and Demand Fully Annotated Graph

5 Creating and Saving Graphs

Naming a Graph

The last graphic function allows you to keep several graphs active in a saved spreadsheet. Press <return> to view the main Graph menu. Next to the Quit option on the right end of the menu is the **Name** function. Once a graph has been created, it can be kept active by entering

/NCname<return>   (name = assigned file name)

This process assigns a name to the graph for future recall *only* if the spreadsheet is saved after the Name command is executed.

When a spreadsheet is saved without resetting, the most recent graph is saved with it, and the graph becomes part of the spreadsheet. The Name command allows more than one graph to be retained. Figure 5.10 shows the **/Graph Name** command structure.

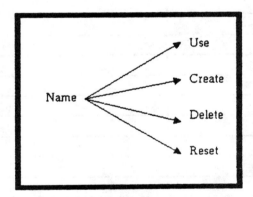

Figure 5.10 File Name Command Structure

If only one graph is to be retained as part of a spreadsheet, there is no need to name it; merely save the spreadsheet. If two or more graphs are to be retained, then enter the following:

NCfilename<return>   (Name, Create, assigned file name)

Creating and Saving Graphs 5

As with other Name menus in 1-2-3, the command line will present all existing names for automatic selection with the cursor. Executing **Create** saves the graph and its settings in the spreadsheet's buffer. This procedure allows the next graph(s) to be entered and allows the graph to be saved for future reference.

When a spreadsheet is recalled and the active graph is to be changed, the **Name Use** command is executed. Again, the command line provides the list of current graphs for automatic selection with the cursor. Changes can be made to any of the named graphs, either manually or through a macro. The tricky part is that *two* steps must be taken if these changes are to be saved. First, the graph must be re-**(C)reated**; permanent changes are buffered through the **Create** command. Second, as always, the spreadsheet must be saved.

The **Delete** command allows a created graph to be deleted from the graph buffer. This feature is helpful because excess information in the spreadsheet (including graphs) fills up RAM, restricts spreadsheet size, and significantly increases the time required for file access. The **Reset** command deletes all active graphs.

5 Creating and Saving Graphs

SUMMARY

- The **Graph Save** command does *not* save the graph; rather, it creates a printable picture file.

- The **Name Create** command stores the named graph in a buffer to be saved with the graph when a /FS (File Save) command is executed.

- The **Name Use** command switches between graphs saved with a spreadsheet.

- The X-axis on line graphs is not to scale.

- Multiple graphs can be saved as active graphs using the **Graph Name** command.

- Five types of graphs may be created: Line, Bar, Stacked Bar, X-Y, and Pie.

- The **Reset** command cancels settings on a single range or the whole graph.

- Lotus allows automatic or manual scale-setting.

EXERCISES

Case Study Two: Conclusion

In Exercise 2.5, the sales information for XYZ Inc. was extracted to a file named XYZEXTR. This file was then expanded to include a summary table. The following exercises can be completed using the spreadsheet in file XYZEXTR.

1. Plot a Min-Max-Avg graph showing the highest, lowest, and average year-to-date sales by sales office. Label the graph and axis, and add a legend. Name the graph ONE.

2. Remove the legend and replace it with line labels on the face of the graph. Name the graph TWO.

3. Generate a pie chart of YTD sales by office. Explode the East office. Name the graph THREE. Explode the West office. Then explode the Midwest office.

4. Plot the year-to-date sales for all of the salespersons by seniority. Sort the data field and replot as sales by rank. Change both of these graphs to bar graphs and replot.

5. Generate a stacked bar graph showing inside, outside, and total sales by sales office. Annotate the graph and name it FOUR.

6. Save the spreadsheet. Reload it and view each of the graphs sequentially.

Answers

1. The first three commands that follow retrieve the spreadsheet and insert two columns with width one. The purpose of these columns is to allow the graph ranges to start and end on a blank cell, which greatly improves the look of line graphs with small numbers of points:

   /frXYZEXTR<return>   File Retrieve
   <F5>l1<return>   GoTo cell L1
   /wic<return>/wcs1<return>   Insert Col, set
                               width 1

   <F5>p1<return>   GoTo P1
   /wic<return>/wcs1<return>   Insert col, set
                               width 1

5 Creating and Saving Graphs

The next group of commands enters Graph mode, resets all graph settings (/grg), and sets the ranges for X, A, B, and C. In each range setting, the inserted columns, N and P, are included:

/grg	Graph Reset
	Graph
x13.p3<return>	Set X range
a14.p4<return>	Set A range
b15.p5<return>	Set B range
c16.p6<return>	Set C range

The next four lines enter the title settings to label the graph and axis:

otfMin-Max-Average Graph<return>	Option Titles First
TSExercise5_1<return>	Titles Second
txSales Office<return>	Titles X-Axis
tySales (units)<return>	Titles Y-axis

Finally, these last three lines provide a legend below the graph:

laMin<return>	Legend A
lbAvg<return>	Legend B
lcMax<return> q	Legend C Quit

And the graph is named as follows:

ncONE<return> v	Name Create "ONE"
	View

Figure 5.11 shows the completed graph.

Creating and Saving Graphs 5

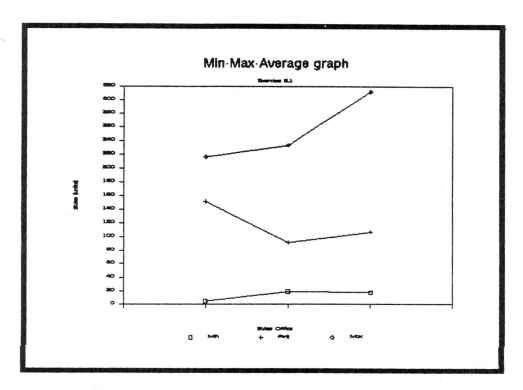

Figure 5.11 Min-Avg-Max Graph

2. Addition of on-page line labels requires creation of data labels. The three cells that follow must be entered to provide data for the label command:

    O21: 'MAX
    O22: 'AVG
    O23: 'MIN

    Once the label data are entered, reenter Graph mode and update. The legends must be reset:

    /gola{esc}<return>   Graph Options Legends A
    lb{esc}<return>   Legends B
    lc{esc}<return>   Legends C

5 Creating and Saving Graphs

The data-labels just entered must be called out in data-label range definition statements. The following statements define the data-labels as the information entered in the previous O column. Note that the range for labels includes the two blank columns just as the data-ranges contained two blank columns:

da121.p21<return>r	Data-labels A Right
db122.p22<return>r	Data-labels B Right
dc123.p23<return>r	Data-labels C Right

The next two lines change the title of the graph and change the display format. The format change removes the symbols from the graph because the legend will be to the right of the third entry:

tsExercise 5.2<return>	Set Second title line
falblcl	Format A Lines B Lines C Lines
qq	Quit Quit

The graph is named TWO:

ncTWO<return>v	Name Create TWO View
q/fs<return>r	Quit File Save Return Replace

Figure 5.12 shows the completed graph.

Creating and Saving Graphs **5**

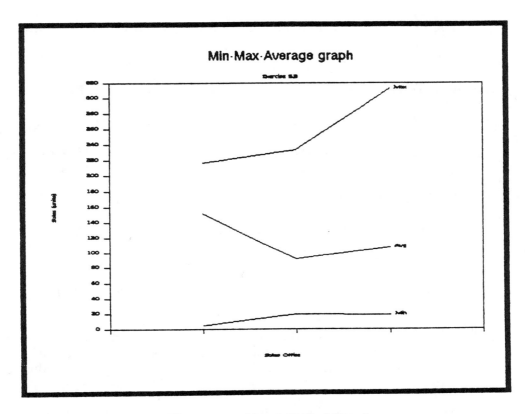

Figure 5.12 Line Labelled Graph

3. Since 1-2-3 does not automatically provide shading, you must specify these values. Acceptable values are 0-7, where 0 is clear. The cell will be exploded if 100 is added to the shade value. The three cells that follow will serve as data for the range statement required to shade this graph:

M24: 101
N24: 2
O24: 3

5 Creating and Saving Graphs

The next four lines generate the graph ranges and titles:

/grgxm3.p3<return>	Graph Reset Graph Set X range
am7.p7<return>tp	A range Type Pie
otfYTD Sales By Office<return>	Options Titles First
tsExercise 5.3<return>q	Titles Second Quit

Setting the B range to the row of numbers entered from M24-O24 tells 1-2-3 what shade and configuration to give the pie slices:

bm24.o24<return>	B range

The graph can now be named and viewed:

ncTHREE<return>v	Name Create THREE View
q/fs<return>r	Quit File Save Return Replace

To explode West, move the cursor to cell M24 and change its value to 1. This change will unexplode East. Change O24 to 103 to explode the West. Press the F10 key to replot the graph on the screen.

To explode the Midwest, set N24 to 102, O24 to 3, and press the F10 key.

Figure 5.13 shows the completed graph.

Creating and Saving Graphs 5

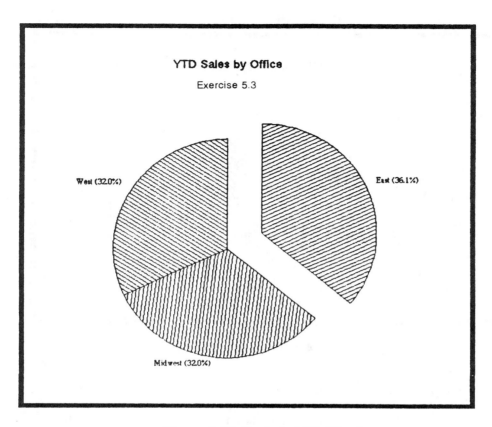

Figure 5.13 Exploded Pie Chart

4. The following commands generate the graph shown in Figure 5.14:

/grgtbxh5.h28<return>	Graph Reset Graph Type Bar X range
ae5.e28<return>	A range
otf YTD Sales by Tenure<return>	Options Titles First
tsExercise 5.4<return>	Titles Second
txSeniority Level <return>	Titles X-Axis
tyYTD Sales (units)<return>	Titles Y
ss5<return>qv	Scale Skip 5 Quit View
q/fs<return>r	Quit File Save Replace

The Scale Skip command limits the number of digits displayed along the X-axis to keep the graph readable.

115

5 Creating and Saving Graphs

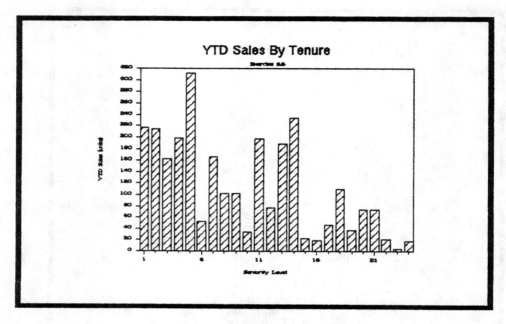

Figure 5.14 Exercise 5.4 Solution

5. The final problem requires the addition of another summary table. Sort the data in this file with Sales Office as the primary key and Position as the secondary key. The following summation statements will build a table of values for the graph. The inside sales row will be the sums of position I salespersons. The outside sales row is position O salespersons plus supervisors and managers:

```
K27:  'Inside
K28:  'Outside
M27:  @sum(e5.e7)
M28:  @sum(e8.e10)
N27:  @sum(e11.e17)
N28:  @sum(e18.e20)
O27:  @sum(e21.e25)
O28:  @sum(e26.e28)
```

Creating and Saving Graphs 5

The following statements generate the required graph:

/grgtsx13.p3\<return\>	Graph Reset Graph Type Stacked-Bar X range
a127.p27\<return\>	A range
b128.p28\<return\>	B range
otfYTD Sales Inside vs Outside\<return\>	Titles First
tsExercise 5.5\<return\>	Titles Second
txSales Office\<return\>	Titles X
tyYTD Sales (units)\<return\>	Titles Y
qnc"FOUR"\<return\>v	Quit Name Create "FOUR" View

q/fs\<return\>r

Figure 5.15 shows the resulting graph. Because the bars are stacked, the total is represented by the top of each bar, and no special range was required to enter that value.

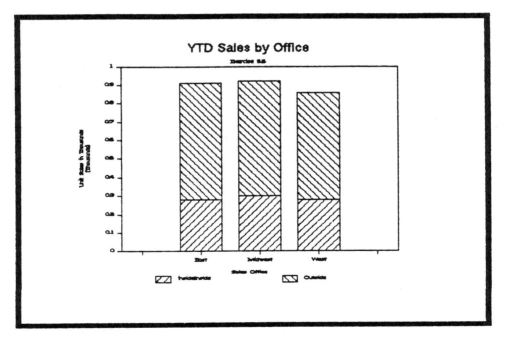

Figure 5.15 Stacked Bar Chart of YTD Sales

5 Creating and Saving Graphs

6. If you are working through these exercises in sequence, the last operation you performed was to save the previous file. If you have not saved it yet, execute a File Save at this time, then execute the following:

/wey	Worksheet Erase Yes
/frXYZEXTR	File Retrieve
/gnuONE\<return>v	Graph Name Use One

    The first graph, named in Exercise #1 of this chapter, will be displayed on the screen. Next enter the following:

\<return>nuTWO\<return>	Name Use TWO

    which will display the second graph. With these commands, any graph named can be recalled. If changes are made, and those changes need to be saved, name the graph a second time using the same name. After this step is done, be sure to execute a File Save.

PART II

SINGLE-VARIABLE ANALYSIS

Single-variable analysis refers to the study of variation of a single parameter. If that parameter is height, for example, you would consider the range of heights of individuals in a group, the average height of an individual, the standard deviation from the average, and so forth. The task, then, is to take a group of readings or values and extrapolate performance of the population from the data.

A more practical example might be fuel economy as considered by an automobile manufacturer. Law requires that a mileage rating be provided for every new car sold. Obviously, no manufacturer will test every car for its mileage. A group of prototypes is evaluated, and the EPA rating is extracted from this data. The techniques presented here will allow you to make projections of this type.

The size and type of the data set play a significant role in determining what type of analysis should be made. If a large data set is available, descriptive techniques exist that allow accurate projections about the population. For smaller groups of data, other techniques are used. Frequently, these small sample techniques — or variations of them — are applied to production lines and process environments to determine if changes are occurring.

Chapter 6, "File Management," provides the first of the templates to be developed in this book. This template allows for input of data sets, import of numerical data from ASCII files, and creation of spreadsheet data files for use by the statistical templates. It is possible to input data directly to the statistical templates developed in this text, but much greater flexibility is allowed by storing data — and operating on that data — separately.

Chapter 7, "Small Sample Statistics," provides methods of analysis suited to small data sets or multiple small data sets. Included here are minimum, maximum, average, median, mode, and Pearsonian skewness. Often, these techniques are applied in a production environment where a few items are tested periodically to see if a process has remained in control. These techniques have the advantages of speed and low cost but lack the accuracy of more comprehensive test methods.

Chapter 8, "Descriptive Statistics," is the heart of this book. These methods apply to normal (or **Gaussian**) data sets and provide more information per dollar of test or analysis time than any other methods presented in this book. Included are mean, deviation, variance, kurtosis, skewness, and histogramming. In the previously mentioned height example, projections about the height of men in general can be made from the mean and deviation of heights taken from a group of men. The histogram and moments (kurtosis and skewness) provide insight into the distribution of the sample.

Chapter 9, "Graphic Methods," provides some techniques for generating graphs to enhance the descriptive statistics from Chapter 8. A line graph of the raw data set can be plotted, as can an **ogive** (cumulative frequency distribution). A table of expected values can be generated based on the mean and deviation of the data set. This table can be plotted concurrently with the histogram from Chapter 9. Lastly, the expected values can be used to perform a **chi-squared** test to evaluate whether the data are from a normally distributed population.

CHAPTER 6

FILE MANAGEMENT

6 File Management

If the user has a single purpose in mind, a spreadsheet can be created in such a way that data can be entered and saved with it. For one-time applications, this type of spreadsheet proves to be the most efficient use of time. If a spreadsheet is to be used repeatedly for the same purpose, it can be stored without the data saved in it. Retrieving a **template** created this way allows rapid entry of data and analysis. If several such templates are available, and more than one template may be applied to a data set, creation of separate data files is logical.

The template presented in this chapter addresses that application. The four chapters in Part II all require a data set that is a single column of values. In Part III, the templates require multiple data sets. To ease documentation of the analysis performed, information about the data set can be stored along with the data set itself. This template should perform the following functions:

- accept the data from the keyboard or an imported file

- provide a means to store it for use by other templates

- allow editing of the data set

- allow existing data sets to be recalled and modified

Using a template is different from using a dedicated (or comprehensive) software package. In a dedicated software package, the operation must be geared to a low-level user. All functions must be automated, and error trapping must be included. In a spreadsheet template, however, the operator may be called on to perform some operations manually. The purpose of this chapter is to simplify and automate routine operations. In the process, an environment is created that is conducive to off-loading some of the spreadsheet manipulations to an assistant. Keep a proper perspective; a template is more complicated than canned software, but it is also far superior in its flexibility.

File Management 6

METHOD

The template created here—named DATAIN on the companion disk—contains a single-level menu. When the spreadsheet is loaded, a marquis is presented that identifies the DATAIN module to the operator. Along with the marquis, the menu is displayed. Figure 6.1 displays the screen as it should appear after the DATAIN spreadsheet is loaded.

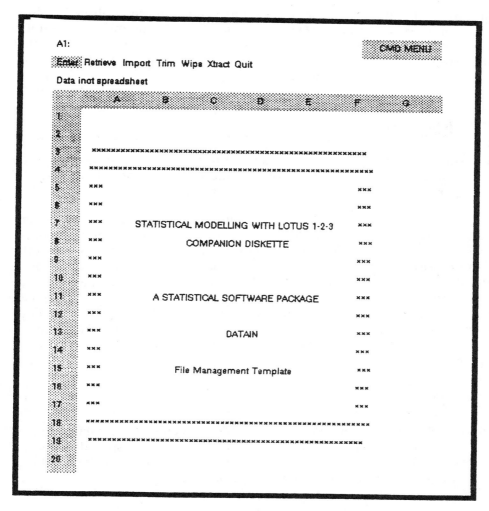

Figure 6.1 DATAIN Template Marquis and Menu

Each function in the menu has an associated submacro that performs the operation and returns control to the menu. The only manual operation is the actual entry of data, which is performed in exactly the same manner as data entry into a standard spreadsheet. After the macro is halted, you move down the data column with the cursor keys, entering data as you go. After data are entered, the macro is restarted, and operations are menu-driven again.

Data entry operation could be fully menu-driven and accomplished within a macro, but, as was stated previously, that is not the intention in template operations. Templates are an aid in mechanizing routine operations; their purpose is not full automation. Details on performing menu operations in 1-2-3 are available in Appendix D.

APPLICATION

The template developed here creates files for data sets consisting of collected groups of single values. These groups could be heights of individuals, EPA mileage ratings, or any other single-parameter data set.

Along with the raw data values, numerous other pieces of information are stored. The data dimension, which is always 1 for one-way analysis, is stored first. Next are the data file name, data description, and name of the operator taking or storing the data. Information about the variable is stored next: variable name, units, upper and lower limits, and count of the data set. Finally, some information to be used in preparing graphs is included. The Title lines (1st and 2nd), X-title, Y-title, and a **floating label** are stored. The floating label is a short description used to highlight graph or spreadsheet ranges in order to distinguish multiple groups from similar data sets.

OUTPUT

The output from this template is a **data file**. Specifically, the data file is itself a spreadsheet suitable for combining with other spreadsheets and templates. At this point, it contains a single column of data and labels. Later, a second column of data values may be added.

File Management 6

REQUIRED COMMANDS

The following commands are used in this macro:

/C	Copy
/FCC	File Combine Copy
/RE	Range Erase
/XC	Submacro Call
/XG	Macro GoTo
/XM	Macro Menu
/XQ	Quit Macro
/WT	Worksheet Titles

TEMPLATE ORGANIZATION

The block diagram in Figure 6.2 illustrates how the template is organized in the spreadsheet. The actual sheet is two screens wide with all data and screen activities on the left in columns A-H. The marquis and all macros are on the right in columns I-O.

6 File Management

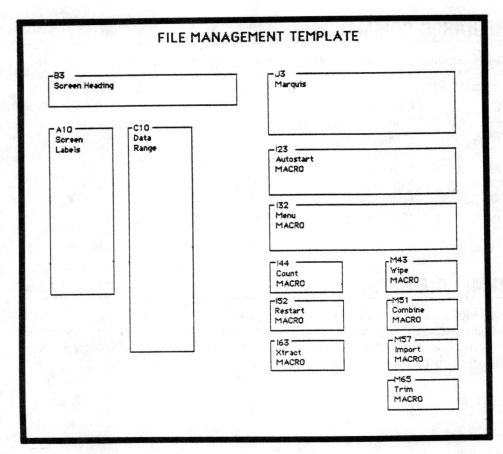

Figure 6.2 DATAIN Macro Layout

The screen heading, marquis, and labels sections all contain labels only. They allow the screen to be descriptive when the macro is executed and minimize the need for operators to work from notes or manuals.

The autostart macro runs immediately when the spreadsheet is loaded. The marquis appears on the screen. The menu macro displays the functional menu in the command line and directs flow to the user-specified macro. The remaining macros are individual, functional routines that perform the specified tasks. Each macro returns control to the DATAIN menu, with the exception of the Quit macro, which calls the "menu" spreadsheet (this process is outlined in detail in Appendix D). The menu spreadsheet is essentially a spreadsheet that, based on user input, can call any of the templates described in the text. This spreadsheet is included on the companion disk and allows the set of templates to be used as a statistics package.

File Management **6**

The next sections describe in detail each of the blocks in Figure 6.2.

MENU

A single-level menu is used in this first template. Figure 6.1 displayed the screen as it appears when the file is loaded. Included in the figure is the menu shown on the command line. Figure 6.3 shows the relationship between the autostart macro and the commands in this menu. Later sections contain multiple-level menus, which will be presented in this same way.

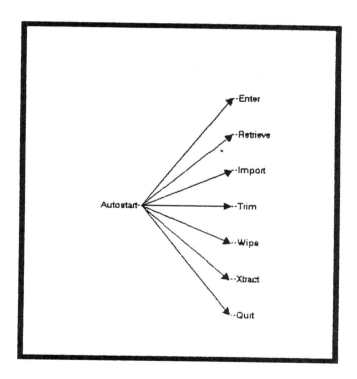

Figure 6.3 DATAIN Template Menu Tree

129

6 File Management

MACRO FUNCTIONS

Variable Labels

The first list of spreadsheet statements is actually not a macro. Each of the cells prints labels (on the screen or printer) that identify data to be entered. For convenience — and because this template is macro driven — the cells will be described here with the macros. Similar sections in later chapters will include calculations.

Figure 6.4 illustrates the screen layout for the DATAIN template. The listing that follows generates the data label section of the screen display. The four sections in a data file are identified in column A: file information, variable information, graph information, and raw data. The name for each variable that can be entered is given in column B. These names can be changed to suit the application. Storing blank spaces is acceptable if data are not needed or available.

File Management 6

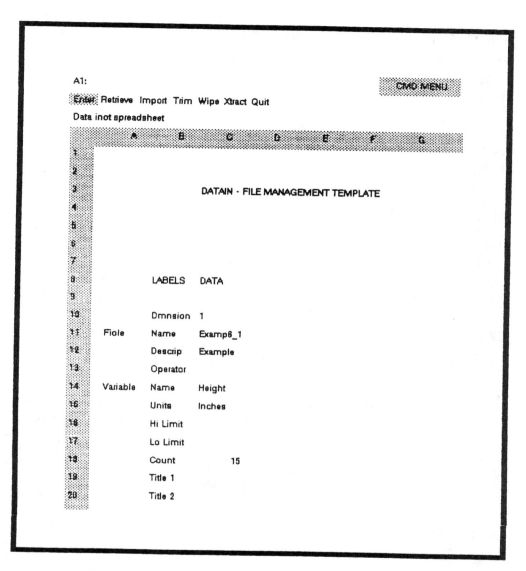

Figure 6.4 Template Display after Restart

6 File Management

Screen Labels and Marquis Listing

```
A11:  'File
A14:  'Variable
A19:  'Graph
A24:  'Data
B8:   'LABELS
B10:  'Dmnsion
B11:  'Name
B12:  'Descrip
B13:  'Operator
B14:  'Name
B15:  'Units
B16:  'Lo Limit
B17:  'Hi Limit
B18:  'Count
B19:  'Title 1
B20:  'Title 2
B21:  'X-title
B22:  'Y-title
B23:  'Flt Lbl
B24:  'Start
C3:   'DATAIN - FILE MANAGEMENT TEMPLATE
C8:   'DATA
I70:  1
J3:   '****************************************************************
J4:   '****************************************************************
J5:   '***                                                          ***
J6:   '***                                                          ***
J7:   '***       STATISTICAL MODELING WITH LOTUS 1-2-3              ***
J8:   '***              COMPANION DISKETTE                          ***
J9:   '***                                                          ***
J10:  '***                                                          ***
J11:  '***          A STATISTICAL SOFTWARE PACKAGE                  ***
J12:  '***                                                          ***
J13:  '***                                                          ***
J14:  '***                                                          ***
J15:  '***              File Management Utility                     ***
J16:  '***                                                          ***
J17:  '***                                                          ***
J18:  '****************************************************************
J19:  '****************************************************************
```

File Management 6

File information consists of the data file name, a description of its contents, the number of data dimensions (1 or 2), and the operator's name. Variable information is the variable name, units, high and low limits on the data set, and the number of data values (computed by the spreadsheet). Graph information consists of two title lines, axis labels, and a floating label for the data set, which will be explained later.

The marquis listing is included as part of the data labels section, which can be deleted if desired. A more suitable approach would be to modify the marquis to suit the application (individual, school, business, etc). This customization to the application gives a professional touch. Simply replace the descriptive lines in the marquis with the name of the individual or organization responsible for the data.

The Autostart Macro

Whenever a spreadsheet is loaded, the autostart macro (always named \0) is executed. The templates in this book have two names assigned to the autostart macros. This allows operation to start automatically, or the user can restart if the macro is accidently halted. The second name is \A. (You *are* allowed to give a single range more than one name.)

The autostart macro that follows moves the cursor Home and then to cell I1 (cells I24-I25). Cell I26 is an input statement that holds the marquis on the screen until return is pressed. Cell I27 puts the cursor in place for running the template. Cell I28 calls the menu.

Autostart Macro Template

```
I21:    'MACROs
I23:    'start
I24:    '{home}
I25:    '{goto}i1~
I26:    '/xlPress return to continue~K23~
I27:    '{home}{goto}c10~/wtb~
I28:    '/xmmenu1~
```

6 File Management

Macro Menu

The next listing includes the macro segments that make up the menu. There are seven options available on startup, each corresponding to one of the submacros that follows. The menu consists of three rows. Row 33 from column I to column O contains the actual commands as they appear on the command line. Row 34 contains the descriptions that appear in the second row of the command line as the cursor scrolls across line one. The third row contains the first line of each macro, initiated by executing the particular menu selection. In this case, each option transfers control to one of the submacros, with the exception of Enter in column I. The Enter macro continues down the I column to I39. Each column could have had a macro listed directly below it, but they are all separated here for the sake of readability. For more details on how menus work, refer to Appendix D.

Macro Menu Listing

```
I32:  'menu
I33:  'ENTER
I34:  'Data into spreadsheet
I35:  '/wtc~
I36:  '{HOME}{goto}b5~
I37:  '    Press alt R when data entry is complete~
I38:  '{goto}c10~
I39:  '/wtb
I40:  '/xQ
J33:  'RETRIEVE
J34:  'Reenter an existing data file
J35:  '/xg\c~
K33:  'IMPORT
K34:  'Data values from a data file (ASCI, ext=.PRN)
K35:  '/xg\i~
L33:  'TRIM
L34:  'Data values from active file leaving settings data
L35:  '/xg\t~
M33:  'WIPE
M34:  'Erase active data completely
M35:  '/xg\W~
N33:  'XTRACT
N34:  'Active data to a file
N35:  '/xg\x~
O33:  'QUIT
O34:  'Return to main menu
O35:  '/frmenu~
```

SUBMACRO OPERATION

There is no difference between a **macro** and a **submacro**. Each is a list of commands that are executed top-down until the end of the list is reached or control is transferred. The term submacro denotes a macro that is used by another macro to accomplish a specific task.

Extract

The following eXtract macro is very similar in nature to the **eXtract** command in 1-2-3; it prints a portion of the spreadsheet into a distinct and separate spreadsheet for later use. In this case, the macro is to be combined with a spreadsheet template and function as the data portion. The extracted portion is itself a spreadsheet.

Cell I45 issues a Home command, which moves the cursor to the top of the data area. The first cell in the row is copied to cell I70, which is an empty cell used to hold this value. It is important to understand why this step is necessary.

In cell I63, after the range to be extracted is entered, one of two conditions occurs. A file **eXtract** command has been issued for the name specified by the operator. Either this file already exists, or it doesn't. If it already exists, the command line will ask if the user wants to replace the file or cancel the request. To make this operation work in the macro, it is necessary to issue the R˜ command, which allows the existing file to be replaced. If the file does not exist, the R value is written to the Home cell, overwriting the dimension (1) that is there. Because this dimension value is stored, it can be recalled from cell I70 later (see I51). It is easier to copy I70 into C10 (the cursor location) every time than to test whether the value had been erased. No harm is done if a "1" is overwritten by a 1.

Cell I47 calls the Count Sum macro (described in the next section). Cell I48 returns the cursor once more to C10 and with the values option, issues the file **eXtract** command. The question mark (?) instructs the macro to expect a keyboard input—in this case, the file name. The {end}{down}{right}{end}{down} in cell I50 prints all data in the column. Trailing blank cells (below the bottom cell in the range) are not stored as blanks. The "r" verifies overwriting of the file if it already exists, as was described previously. The Copy command in cell I51 brings the dimension value back from the cell where it was stored. Control is returned to the menu by the Menu command (/XM) in cell I52.

6 File Management

Extract Macro

```
I43:    'EXTRACT
I45:    '{home}
I46:    '/c~170~
I47:    'xccount~
I48:    '{home}/fxv{?}~
I49:    '.{pgdn}{down}{down}{down}
I50:    '{end}{down}{right}{end}{down}~r~
I51:    'ci70~~
I52:    '/xmmenu1~
I54:    'Count
I55:    '{goto}c18~
I56:    '@count(
I57:    '{down}{down}{down}
I58:    '{down}{down}{down}
I59:    '.{end}{down})~
I60:    '/xr
```

Count

The eXtract macro just listed calls this next submacro to verify and store the number of data values in the raw data range. The cursor is moved to cell C18, where the value is to be stored by cell I56. The @Count command is issued at this point. Cells I57 and I58 move the cursor to the beginning cell in the raw data range. This step is required — as opposed to entering the actual cell location — so that the {end}{down} sequence can be used in cell I59. After the period anchors the range, this step moves the cursor to the last cell in the data range. If an actual cell location had been entered, instead of moving the cursor to the initial cell, the {end}{down} sequence would have caused an error. The closed parenthesis [)] and tilde (~) in cell I59 finish the count routine, and cell I60 executes the subroutine's return to the calling location.

Count Submacro

```
I54:  'Count
I55:  '{goto}c18~
I56:  '@count(
I57:  '{down}{down}{down}
I58:  '{down}{down}{down}
I59:  '.{end}{down})~
I60:  '/xr
```

Restart

Restart is required whenever the macro is halted for entering data or modifying an existing file. The first cell erases the restart message from cell B5. Cell I64 sends the cursor to cell C18 to update the count. Cell I65 calls the Count submacro. I66 returns the cursor to the top of the data range. The final cell returns control to the menu.

Restart Macro

```
I62:  'restart \R
I63:  '/wtc~{home}{goto}b5~/re~~
I64:  '{goto}c18~
I65:  '/xccount~
I66:  '{goto}c10~
I67:  '/wtb~
I68:  '/xmmenu1~
```

Erase

Two macros exist for erasing data from the spreadsheet. The **Wipe** routine erases all information from the top of the column to the end of the data row. The **Trim** routine starts at the top of the raw data section and eliminates all raw data. The two routines are used similarly.

6 File Management

Wipe is used before recalling a file from disk to make sure the range is empty. An error can occur if an old file is overwritten by new data because empty cells in the new file do not overwrite old values. This command also empties the column in preparation for entry of new data.

Trim allows only raw data values to be removed. Frequently, groups of data with similar graph and limit settings will be entered. Erasing only the raw data values makes creating groups of files easier because common information need not be re-entered each time.

Each macro consists of a Range Erase followed by a Return to the macro menu. It is acceptable to erase a range from C10 to the end of the spreadsheet. This approach is easier than locating the actual end point.

Wipe Macro

```
M43:  'WIPE
M45:  'Erases labels and vals
M46:  '{home}
M47:  '/re.{pgdn}{down}{down}{down}
M48:  '{end}{down}~
M49:  '/xmmenu1~
```

Trim Macro

```
M65:  'TRIM
M66:  'Erases vals only
M67:  '{goto}c24~
M68:  '/re.{end}{down}~
M69:  '{home}
M70:  '/xmmenu1~
```

Combine

The **Combine** macro allows an old data file to be recovered from disk. There are numerous reasons why this recovery is valuable. Data can be added or corrected, or major changes can be made in the file settings. Data can be restored in the same file or in a new file created from this point. To do this, move the cursor to C10 (cell M53) and execute a File Combine Copy Entire command. Remember, the question mark (?) causes 1-2-3 to wait for you to enter the file name.

Combine Macro

```
M51:  'COMBINE
M52:  'Reenter existing data file
M53:  '{home}
M54:  '/fcce{?}~
M55:  '/xmmenu1~
```

Import

The final macro in this set is an import routine. Import differs from Combine in that the file retrieved from disk is not a spreadsheet but an ASCII file. Some manual intervention may be required to make the values suitable for the spreadsheet.

Only raw data values can be retrieved using this Import macro. M59 moves the cursor to the top of the raw data range. M60 executes a File Import Number and waits for a file name to be entered. Values are placed in the raw data range.

Import Macro

```
M57:  'IMPORT
M58:  'Enter data values only
M59:  '{goto}c24~
M60:  '/fin{?}~
M61:  '{home}
M62:  '/xmmenu1~
```

6 File Management

CELL RANGE NAMES

If you are entering the macros in this chapter from the text, you must enter the range names for the macros to function. The following commands will accomplish all the necessary range name creations:

```
/RNC\A~I24~                 Manual Autostart
/RNC\O~I24~                 Autostart
/RNC\I~M58~                 Import
/RNC\R~I52~                 Restart
/RNC\T~M66~                 Trim
/RNC\W~M45~                 Wipe
/RNC\X~I61~                 eXtract
/RNC\COUNT~I44~             Count
/RNC\C~M52~                 Combine
/RNC\Menu1~I33.)34~         Menu1
```

PUTTING THEM TOGETHER

Now that all the segments have been entered, the program can be run. If this template is the first one you have entered, load it directly with the File Retrieve command (/FR DATAIN <return>). On the companion disk, the spreadsheet is stored as DATAIN. As soon as the spreadsheet loads, the marquis appears on the screen along with the command menu. The menu appears in the command window at the top, and execution halts, waiting for user input.

File Management 6

The first user option is **Enter** to enter data. Execution of the macro is terminated while the user enters data values in column C; then, it is restarted by the <Alt/R> command. While the cursor is on the **Enter** option, press the Return key or <E>. The cursor keys can then be used to move the cursor. Move the cursor down the row and enter the following:

Enter	At
1	Dimension
Examp6-1	File Name
Example	Descrip
	Operator
Height	Variable Name
Inches	Units
	Lo Limit
	Hi Limit
	Count
	Title 1
	Title 2
	X-title
	Y-title
Examp	Flt Lbl
67	Start
71	
73	
68	
67	
66	
67	
69	
70	
71	
73	
74	
73	
71	
72	
<alt/R>	

141

6 File Management

After these commands have been entered, the <Alt/R> command restarts the macro, and the macro waits for an additional command.

This file can now be saved to disk by executing the **eXtract** command. Either move the cursor to **eXtract**, or press <X>. When the prompt appears, enter

```
examp6_1<return>
```

This command will write the file to disk for later use.

Next, enter the **Trim** command to erase the data values from the bottom of the column. If the screen action is too fast to verify correct operation, use the **Enter** option again. This option will halt the macro and allow you to move the cursor around to inspect the data zones. <Alt/R> restarts the macro. After execution of the **Trim** command, you will be able to scroll down with the cursor keys and verify that all of the raw data values have been erased. The upper part of the column will be left intact.

Press <Alt/R> again, and execute the **Wipe** command. This command erases the whole data group, including the settings. At this point, the column should be clean and the cursor at the top of the screen.

Retrieve, the second command, prompts you for a file name and then automatically loads that file into the data area. Because the **Combine** command is used, this area should be erased with a **Wipe** command first, as was done in the previous step. Move the cursor to the **Retrieve** option, and press the Return key or press <R>. Using the cursor on the command line, move to "examp6_1" and press the Return key. The previously stored file will be loaded into the data column. It can be edited, trimmed, wiped, or just reviewed.

The **Import** command allows you to load raw data values while keeping the data settings in the top portion of the data area. This command is especially handy if data files are created by an external program and most of the settings do not change much from file to file. If minor changes are needed, use the Enter option to edit the data. No Import example is presented here because a suitable data file must first be prepared. Refer to Appendix C for more information on this command.

File Management **6**

Quit executes a /frmenu˜, leaving DATAIN and reloading the main menu spreadsheet. If this spreadsheet is not on your disk, an error will occur. Refer to Appendix D for details on entering and saving the main menu spreadsheet to allow interaction of the templates in this book.

Each of these routines — with the exception of Quit — transfers control back to the menu to continue execution in macro mode. Files created with this template will function for all templates in Part II of this book. In Part III, additions will be made to allow entry of multiple-column data.

SUMMARY

- Data files can be created with the DATAIN template in the form of spreadsheets.

- Files can be reloaded and edited later.

- The Trim and Wipe commands allow removal of information from the data column.

- Macro and subroutine operations require names for certain ranges on the spreadsheet.

EXERCISES

1. Enter the data in the following table into two data files named "exer6_1" and "exer6_2."

File Name	exer6_1a	exer6_1b
Description	Sample Data	Sample Data
Operator	Sean C	Sean C
Data	1	6
	2	7
	3	8
	4	9
	5	10

   Table 6.1 Data for Exercises

6 File Management

2. Merge the two data files created in Exercise #1 into a single file with the name "exer6_1." The resulting file will have a single column of data containing all values from both files in the raw data section.

   Hint: Retrieve "examp6_2"; use (E)nter to stop the macro, and manually move the data to column c. Retrieve "examp6_1" and manually move the data to make one column. Restart the macro and save as "examp6_1."

Answers

1. Load 1-2-3 and retrieve the template stored in the file DATAIN. After 1-2-3 loads, the menu for data entry will automatically appear in the command line. Execute (E)nter:

E	Cursor to C10
	Restart message displayed

   The data from the first column in Table 6.1 is entered using the following commands:

1{down}	Dimension
Exer6_1a	File Name
Sample Data{down}	File Description
Sean C.{down}	Operator
<F5>C24<return>	GoTo C24
1{down}	
2{down}	
3{down}	
4{down}	
5{down}	Data Values Entered
<Alt/R>	Count Entered
	Macro Restarted
xexer6_1a<return>	Data Extracted to File
t	Trim

File Management 6

After the previous commands have been entered, the first data file will have been created. The data values will have been erased by the **Trim** command, and the menu will still be active in the command line. Press (E)nter and repeat the previous process for the second column in Table 6.1. Since the **Trim** command was used, only the data element portion of the file was erased. The dimension, description, and operator name do not need to be re-entered. Enter the new file name in C11; then, go to C24 to enter the data. Substitute the second file name, "exer6__1b," at the eXtract command the second time through the process.

When this process is complete, execute a **Wipe** command to erase the data.

2. With the DATAIN macro active, execute a (R)etrieve:

   rexer6_1a<return>   Combine Copy Entire
                                        executed by the macro.
   E   Enter mode executed.
                                        Macro Halted.

   /m.<pgdn><pgdn><return>
   <right><right><return>   Data moved from Col C
                                        to E.
   <Alt/R>   Macro Restarted.

   At this point, the first data file has been moved so that the second data file can be retrieved. The macro has been restarted so that the Retrieve command can be executed a second time:

   rexer6_1b<return>   Second data file loaded.
   E   Enter mode halts macro.
   /ree10.e23<return>   Remove labels from first
                                        data file.
   /me24.e28<return>c29<return>   Move data elements from
                                        file 1 to cell below last data
                                        element in file 2.
   <Alt/R>   Restart macro.
                                        Count elements.
                                        Menu active.
   xexer6_1B<return>w   Final data file extracted.
                                        Wipe executed to erase
                                        data range.

145

6 File Management

TEMPLATE LISTING

The following listing was created by transferring a working copy of the spreadsheet into a word processor. If you are entering the spreadsheet manually, use this copy instead of the segments within the chapter. Those segments are provided for instructional purposes only and may differ slightly from this complete listing.

```
A11:  'File
A14:  'Variable
A19:  'Graph
A24:  'Data
B8:   'LABELS
B10:  'Dmnsion
B11:  'Name
B12:  'Descrip
B13:  'Operator
B14:  'Name
B15:  'Units
B16:  'Lo Limit
B17:  'Hi Limit
B18:  'Count
B19:  'Title 1
B20:  'Title 2
B21:  'X-title
B22:  'Y-title
B23:  'Flt Lbl
B24:  'Start
C3:   'DATAIN - FILE MANAGEMENT TEMPLATE
C8:   'DATA
I70:  1
```

continued...

...from previous page

```
J3:  '************************************************************
J4:  '************************************************************
J5:  '***                                                      ***
J6:  '***                                                      ***
J7:  '***        STATISTICAL MODELING WITH LOTUS 1-2-3         ***
J8:  '***                 COMPANION DISKETTE                   ***
J9:  '***                                                      ***
J10: '***                                                      ***
J11: '***           A STATISTICAL SOFTWARE PACKAGE             ***
J12: '***                                                      ***
J13: '***                                                      ***
J14: '***                                                      ***
J15: '***              File Management Utility                 ***
J16: '***                                                      ***
J17: '***                                                      ***
J18: '************************************************************
J19: '************************************************************
```

Marquis and Labels

```
I21: 'MACROs
I23: 'start
I24: '{home}
I25: '{goto}i1~
I26: 'xlPress return to continue~k23~
I27: '{home}{goto}c10~/wtb~
I28: '/xmmenu1~
```

Autostart Macro

```
I32: 'menu
I33: 'ENTER
I34: 'Data into spreadsheet
I35: '/wtc~
I36: '{HOME}{goto}b5~
I37: '    Press alt R when data entry is complete~
I38: '{goto}c10~
I39: '/wtb
I40: '/xQ
J33: 'RETRIEVE
```

continued...

6 File Management

...from previous page

```
J34:  'Reenter an existing data file
J35:  '/xg\c~
K33:  'IMPORT
K34:  'Data values from a data file (ASCI, ext=.PRN)
K35:  '/xg\i~
L33:  'TRIM
L34:  'Data values from active file leaving settings data
L35:  '/xg\t~
M33:  'WIPE
M34:  'Erase active data completely
M35:  '/xg\W~
N33:  'XTRACT
N34:  'Active data to a file
N35:  '/xg\x~
O33:  'QUIT
O34:  'Return to main menu
O35:  '/frmenu~
```
                     Menu Macro

```
I43:  'EXTRACT
I45:  '{home}
I46:  '/c~i70~
I47:  '/xccount~
I48:  '{home}fxv{?}~
I47:  '.{pgdn}{down}{down}{down}
I50:  '{end}{down}{right}{end}{down}~r~
I51:  '/ci70~~
I52:  '/xmmenu1~
```
               Extract Macro

```
I54:  'Count
I55:  '{goto}c18~
I56:  '@count(
I57:  '{down}{down}{down}
I58:  '{down}{down}{down}
I59:  '.{end}{down})~
I60:  '/xr
```
                 Count Macro

continued...

148

File Management 6

...from previous page

```
I62:  'restart \R
I63:  '/wtc~{home}{goto}b5~/re~~
I64:  '{goto}c18~
I65:  '/xccount~
I66:  '{goto}c10~
I67:  '/wtb~
I68:  '/xmmenu1~
```
            Restart Macro

```
M43:  'WIPE
M45:  'Erases labels and vals
M46:  '{home}
M47:  '/re.{pgdn}{down}{down}{down}
M48:  '{end}{down}{right}{end}{down}~
M49:  '/xmmenu1~
```
            Wipe Macro

```
M51:  'COMBINE
M52:  'Reenter existing data file
M53:  '{home}
M54:  '/fcce{?}~
M55:  '/xmmenu1~
```
            Combine Macro

```
M57:  'IMPORT
M58:  'Enter data values only
M59:  '{goto}c24~
M60:  '/fin{I}~
M61:  '{home}
M62:  '/xmmenu1~
```
            Import Macro

continued...

6 File Management

...from previous page

```
M65:  'TRIM
M66:  'Erases vals only
M67:  '{goto}c24~
M68:  '/re.{end}{down}{right}{end}{down}~
M69:  '{home}
M70:  '/xmmenu1~
                    Trim Macro
```

CHAPTER 7

SMALL SAMPLE STATISTICS

7 Small Sample Statistics

As you collect data for analysis, carefully consider what techniques will be used to analyze that data. Too often, experimenters and managers wait until a job is done to make this decision. Frequently, extra work must be done to gather information that could have been acquired as a matter of course. One advantage of developing a cohesive software system—like the one described here—is an understanding of the types and quantities of data required. Having this foreknowledge will give you a leg up on success.

Different analytical methods are applied, based on sample size, sample type, and the intended output type. For small sample sizes (usually five to 25 elements per sample), the average and range or minimum and maximum methods are used. The standard deviation can be misleading with small samples, and histograms generally require 30 or more points to be effective. The median and mode, however, can provide information about how skewed the data are even for small data sets.

The first statistical template to be developed—SMALL—allows you to operate on a data set and extract the following information:

- Arithmetic mean
- Count (number of elements in the file)
- Minimum and Maximum
- Median and Mode
- Pearsonian skewness coefficient
- Modality (positive indication of mode is not unique)
- Graph of Min, Max, Average by data group
- Simultaneous analysis of up to six groups
- Printed output report

Input to the SMALL template is entered either from the keyboard or from the DATAIN template created in the previous chapter. Data will be presented on the screen or sent to the printer. The data table allows up to six data files to be compiled into summary form. The physical size of the output report is kept small to allow a graph to be printed on the same sheet.

The actual layout of the template is shown in Figure 7.1. The format is three screens wide, which means that pressing the Tab key twice moves the display window to the right to include the last column. To ease the learning process, the templates developed in later chapters will be kept similar in organization. In this case, the summary area, data range, and marquis are in the same locations as they were in the DATAIN template from Chapter 6.

7 Small Sample Statistics

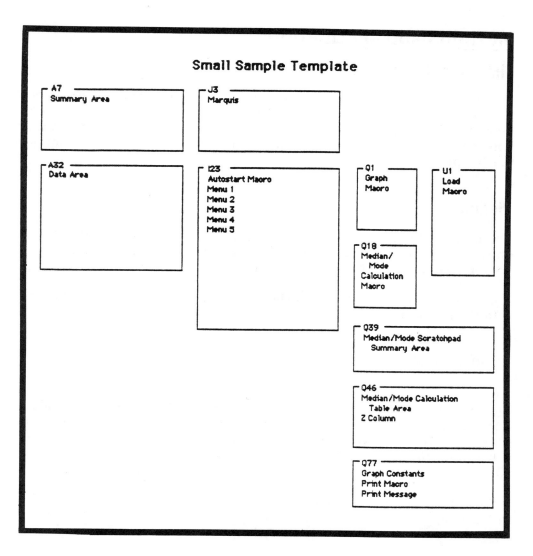

Figure 7.1 Small Macro Layout

This template is much more complex than the data input macro. There are three menus and a scratchpad area. **Scratchpad** is a programmer's term for an area in which calculations can be made. This area is required because no command is available for directly calculating the median or other functions described in this chapter. There must be an area in the spreadsheet in which to make the intermediate calculations and pick off the results.

7 Small Sample Statistics

METHODS AND APPLICATIONS

Min, Max, Average

The **mean**, **count**, **maximum**, and **minimum** are all what they appear to be. Each is directly calculated from the data column using the corresponding @ function. The count is one of the variables stored in the data file, but it is updated here because users will frequently make data changes when running this template.

Median

The **median** is the center value in a data set. Half the values in the set fall on either side of this value. If the values are ages in an average family, figuring the median would be a simple computation.

For example, an average family might include two parents and three children with the following ages:

42, 40, 17, 12, 9

The median age here is 17; two values fall on each side of this median value. For small data sets with a high degree of granularity, the median is more sensitive than the mean. **Granularity** is the measure of how spread out the data values are. In this case, there are no duplicate values; in fact, there are no adjacent values, which is an example of a high degree of granularity.

If the same data were considered with a measurement resolution of decades rather than years, the values would be as follows:

0-9	1
10-19	2
20-29	0
30-39	0
40-49	2

154

Now, instead of five elements with unique values, there are three cells, or ranges, containing the elements. This change represents a reduction in granularity. For clarity, each cell can be represented by a single value, usually the top value. The cell limits here become 9, 19, 29, 39, 49, and so forth. The median is still a point where half of the values fall above and half below it, but it is no longer easy to identify. There is a simple format you can use to find the median:

1. Find the top of the cell below the median (9).

2. Find the percentage of points in the median cell that fall below the median (2 of 4 = 50% or .05).

3. Multiply the percentage of points below the median (.5) by the size of the median cell (10), and add that amount to the top value of the next lowest cell (9) .

$(9 + [0.5*10] = 14)$

The value of 14 found for the median differs from the original value (17). This is not an error; it is a difference in interpretation based on the resolution of measurement and calculation. In the first case, the resolution of both measurement and calculation was years. In the second case, the calculation resolution was reduced to decades, with each decade representing a cell range for one of the ages. This strategy is occasionally referred to as a **cellular calculation** technique.

A second example of calculating a median might be found in exam scores. The following data set contains scores on a high school algebra test:

100, 97, 94, 92, 91, 91, 88, 88, 88, 87, 87, 87, 87, 86, 86, 86, 86, 86, 86, 86, 85, 85, 85, 84, 84, 84, 82, 81, 78, 77, 74

This class of 31 students has a median score of 86. This value is not very accurate, as seven students received a score of 86. Twenty of the 31 students scored between 84 and 88, which shows a low degree of granularity.

7 Small Sample Statistics

The cellular technique can be used here to expand the accuracy of the median calculation. This technique is based on making a frequency distribution table for an intermediate calculation. The cell below the median is identified, and then the reported median is calculated based on the portion of the median cell that is allocated to each side of the distribution. In this example, the median is 86.6; slightly more than half of the scores of 86 were below the median. Final details on how this calculation was performed, and the equation used, are provided when the operation of the macro is explained later in this chapter.

Comparing the results of the two examples illustrates why the SMALL template is more effective with certain types of data sets. Figure 7.2 is an output report from this chapter's template. Data were created using the DATAIN template. In the first example, the mean age of family members is 24, with a median age of 17. The fact that no member of the set was close to the mean illustrates that the data are skewed. **Skewness** is a measure of how far off center (mean) the median data value is. For very small data sets, the difference between average and median can be quite large. Even larger data sets can be affected significantly by a single value. In other words, the mean is sensitive to change by an outlying value.

The high school test score example is a much larger set with few outlying values on each side of the mean. Only .2% difference exists between the mean and median. Little is learned from this calculation. The sample size of the student exam scores example is more suited to larger sample techniques like **variance** and **standard deviation**. These techniques are the subjects of a later chapter.

Small Sample Statistics 7

```
           STATISTICAL MODELLING WITH LOTUS 1-2-3

                 SMALL SAMPLE STATISTICAL ANALYSIS

         GROUP        1         2         3         4         5         6
File              exmp7_1A  exmp7_1b  --------  --------  --------  --------
Description       Example   Example   --------  --------  --------  --------
Operator          Will      Will      --------  --------  --------  --------

Date              06-Feb    06-Feb    --------  --------  --------  --------
Variable          Age       Grade     --------  --------  --------  --------
Units             Years     Percent   --------  --------  --------  --------
Low Limit                             --------  --------  --------  --------
High Limit                            --------  --------  --------  --------
Data Count            5          31   --------  --------  --------  --------

Data Label                  Grde

Minimum               9          74
Maximum              42         100
Average              24    86.38709
% OL High           0.0         0.0   --------  --------  --------  --------
% OL Low          100.0       100.0   --------  --------  --------  --------

Median               18    85.71428   --------  --------  --------  --------
Pears Skew     0.424688    0.130551   --------  --------  --------  --------
Mode                 42          86   --------  --------  --------  --------
Bi-Mode               1           0   --------  --------  --------  --------
```

Figure 7.2 **Sample Output with Two Group Sizes**

Mode

Mode is the value in a distribution that occurs most frequently. The mode calculation is also based on a frequency distribution but is conceptually simpler. After the frequency distribution table is generated, it can be sorted in order of decreasing number of items in each cell. The cell with the largest count is the mode cell. If more than one distinct mode is found, the distribution is possibly bimodal or multimodal. Figure 7.3 shows the distribution of a bimodal data set. The top figure is the distribution of grades in a math class. At first glance, it appears to be fairly normal. The deviations near the peak could be a random sampling error; however, they represent the sum of the two distinct groups shown in the second picture.

7 Small Sample Statistics

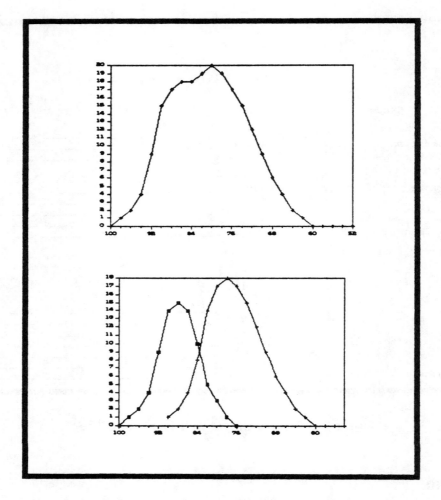

Figure 7.3 Bimodal Distribution Example

Bimodal is a statistical term indicating that more than one distribution is included in a set of data. This situation is clearly evident in the example of ages in a family including two generations. The high school grades could also have been from a bimodal distribution if the class had been part honors students and part regular students.

The bimode indicator in the template developed here is not intended to assert whether or not there is a bimodal distribution. It only indicates that two or more cells tied for the largest number of occurrences. These distributions should be inspected carefully by the analyst. Two adjacent cells can have the same count even in a purely Gaussian distribution.

Another indicator that could have been used in place of—or in conjunction with—the first indicator is separation between peaks. If the cells containing the two or three largest counts are widely separated, it is another indication of a potentially multimodal distribution.

The graph output provided is a Min-Max-Avg graph report. It will contain an entry for each data set in the summary report. Scale will be determined by the overall range of values. This form was chosen for two reasons. The graph is a standard, widely accepted graph type, and, more importantly, it gives a visual and useful description of how close each mean is to the end points of the sets.

Adapting the graph output to average and range (X bar and R) or average and median is easy. With a little more effort, the number of cells can be increased. These applications are quite specific and will be left as exercises for those who need them. Most branch or computational ranges in the spreadsheet macro have been addressed by name to facilitate this operation. The user needs only to adjust references in formulas to specific cell locations that have been moved by the insertion of columns.

For the purposes of this book, data are usually taken in small groups. There are a number of reasons for this strategy: equipment limitations, time limitations, availability of material for testing, and variation from group to group, to name a few. These small groups of data are stored in individual files on disk and summarized using this template. The data can be combined into larger data sets and can be analyzed using more descriptive statistical methods. This combination of applications allows the most complete understanding of the data. Presentation-quality outputs are an additional benefit.

Up to this point, only one-dimensional data sets have been considered. In Part III of this text, **two-way analysis** techniques will be introduced. These techniques operate on pairs of data. At that time, the DATAIN file will be modified to create files with two raw columns. With some limitations, such files can be successfully summarized with this template.

7 Small Sample Statistics

OUTPUT

The outputs provided in this template have been briefly presented. Following is a summary of these outputs:

- half-page summary table
- half-page summary with data table including raw data values
- half-page summary report with Min-Max-Agv graph

REQUIRED COMMANDS

The following commands are used in this template:

/G	Graph commands
/WCS	Worksheet Column-width Set
/WT	Worksheet Titles
/PP	Print to Printer
/XC	Macro Subroutine
/XG	Macro GoTo
/XM	Macro Menu
/Xn	Enter a Number into a cell from a macro
/XR	Return from Subroutine
/FR	File Retrieve
/C	Copy range
/FCC	File Combine Copy
@Today	Print Today's date
@IF	Conditional value placed in a cell

Small Sample Statistics **7**

MENUS

All the templates in this book contain certain similarities. They each open with a marquis describing which template has been loaded. Each also opens with an active menu in the command line at the top of the screen. Layout of the spreadsheets is also consistent where possible. The report or summary area will be in the upper left corner, data will be below, and macros will be to the right of the summary area. This arrangement allows unlimited freedom for entering data values below the report.

It is not necessary to describe the autostart and initial menu operation in each chapter; rather, in this and later chapters, the actual menu structure will be shown. Only macro sections with a unique or complex function will be fully described and listed in the chapters. Each chapter will conclude with the template listing. There may be slight differences between the macro segments within the chapter and the full template listing at the end of each chapter. The full listing is intended for loading and maintaining the spreadsheet. The abbreviated listings within the chapters are for instruction only.

The menu structure for the SMALL template is shown in Figure 7.4. When the spreadsheet is loaded, the cursor will be at the top of the leftmost of the six possible data ranges. From this position, data can be entered into column one. The File Load operation always functions on the column at the current cursor location. Columns should always be added from left to right in this template because skipped columns will block graph action.

7 Small Sample Statistics

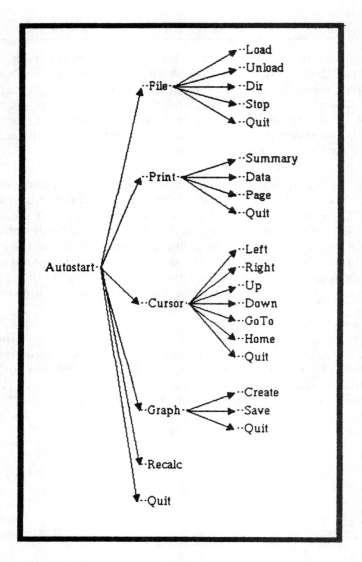

Figure 7.4 Small Template Menu Structure

Main Menu

There are six options available in the main or startup menu. The first four — (F)ile, (P)rint, (C)ursor, and (G)raph — each have submenus. (R)ecalc performs the same calculations made when a file loads. This option allows a user to stop the macro, enter additional data or changes, and recalculate the macro. (Q)uit loads the MENU template, which allows the user to exit the package or link to another function.

File

Entering (F)ile while in the main menu causes the File menu to be executed. The user is optionally allowed to (L)oad or (U)nload a file, change the default (D)irectory to cause 1-2-3 to use a different disk drive for storage of data files, or (S)top macro execution. (Q)uit returns control to the main menu.

Not only does the File Combine function take a long time, but a considerable amount of movement and blinking occurs while the command executes. This process will be thoroughly explained as part of the macro description, but essentially, the command moves data from the input range to the summary range. It also creates a scratchpad, sorts the raw data, and calculates median and mode from the results. Lotus 1-2-3 release 2 users can mask these operations from the screen to speed up operations.

Print

At the conclusion of the File Load command, a Print Summary can be executed. The Print menu is accessed by pressing (P)rint or moving the cursor to the Print command and pressing <return>. The user can print a (S)ummary, list the (D)ata, or execute a form feed with the (P)age command. (Q)uit returns control to the primary menu. The form feed allows the user to output a single report per page. Reports are in a half-page format to allow combination with graphs.

7 Small Sample Statistics

Cursor

The Cursor submenu allows the operator to move the cursor around the spreadsheet to look at values or portions of the software. The Goto command also allows the user to move directly to a spreadsheet location. This step is accomplished by entering (G)oto and a cell location for the address. (H)ome returns the cursor to C10. (Q)uit returns control to the primary menu.

Graph

The Graph menu allows the user to select (C)reate or (S)ave to generate or store a graph. (Q)uit returns control to the primary menu. The Graph function generates the Min-Max-Avg graph based on data in the various columns. Labels for the graph are taken from the first data column.

MACRO DESCRIPTIONS

The data area, marquis, autostart macro, and menu macros all work almost identically to those in the last chapter. Listings for each are included in the template listing at the end of this chapter. The quickest way to find a particular segment in this listing is to look at the layout sheet (Figure 7.1) and find the corner of the range. For example, the Autostart macro starts at cell I22.

Load

The Load macro is executed from the File menu. It actually serves two purposes: it retrieves the data from a disk file specified by the operator and performs the calculations required by linking to the Median/Mode macro below it.

Lines U1-U4 enter the date in line 11 and position the cursor to the first cell in the data range. Line U5 retrieves the file.

Small Sample Statistics 7

U1: 'Load
U2: '{down}{down}{down}{down}
U3: '@today~/rfd2~{pgup}
U4: '{pgdn}{pgdn}{up}{up}{up}
U5: '/fcce{?}~

The next 14 lines move data from the input file to the display area so that data will be part of the output, along with being used in labeling graphs. The data count is transferred to the scratchpad area for later use. Line U16 paints all the imported data elements and names their range as "DATA." (The previous range of the same name is deleted first.) The cursor is finally moved up to the row containing the data count.

U6: '{down}
U7: '/c{down}{down}~
U8: '{pgup}{pgup}{down}{down}~
U9: '{down}{down}{down}/c{down}{down}{down}{down}~
U10: '{pgup}{Pgup}{down}
U11: '{down}{down}{down}~
U12: '{DOWN}{DOWN}/c{down}~u40~

U13: '{DOWN}{DOWN}{DOWN}{down}
U14: '{DOWN}{DOWN}{DOWN}
U15: '/c~{pgup}{pgup}{down}~
U16: '{down}/rnddata~/rncdata~.{end}{down}~
U17: '{pgup}{up}{up}{up}{up}{up}{up}
U18: '{up}{up}{up}{up}{up}
U19: '{up}{up}{up}{up}{up}

Cell U20 finds the number of elements in the data range and enters it in the count line. This value is converted from an equation to a value using the {edit}{calc} sequence. The number is transferred to the scratchpad area. Cell U21 moves the cursor to the minimum line. The next three lines repeat this process for min, max, and average. The final line — U25 — transfers macro control to the MMC macro, where median and mode are found.

7   Small Sample Statistics

Load Macro

```
U20:  '@count(data)~{edit}{calc}~/c~u42~
U21:  '{down}{down}{down}{down}
U22:  '@min(data)~{edit}{calc}~/c~s40~{down}
U23:  '@max(data)~{edit}{calc}~/c~s41~{down}
U24:  '@avg(data)~{edit}{calc}~{down}
U25:  '/xgmmc~
```

Median/Mode

The first group of commands creates a frequency distribution chart for calculation of the median and mode. Q19 and Q20 input and store the upper limit of the first cell and the cell size. The next cell — Q21 — uses these values to fill a data range for the Data Distribution command. U22 generates the distribution based on the current data column and the values selected.

```
Q18:  'Median/Mode calculations
Q19:  '/xnEnter the Top of the lowest cell ~s42~
Q20:  '/xnEnter the cell size ~s43~
Q21:  '/dfbin~s42~s43~~
Q22:  '/dddata~bin~
```

The equations from cells U94 and V94 (Equations 7.1 and 7.2) are copied into the table by cell Q23. These two columns have similar properties when they are entered. Boundaries in column Q are compared to the upper limit specified in the data file. If the cell limit is lower than the specified limit, the @if statement records the contents of column R (the cell count) in column S. In a similar manner, the values are compared to the upper limit and stored in column T. When this action is complete, column S will contain cell counts of cells that are less than the specified minimum, and column T will contain those greater than the specified maximum. Lines Q24 and Q25 sum these two columns and convert the result to a percent for the display.

Small Sample Statistics 7

$$\% \text{ OL Low} = \frac{\left[\sum_{n=LC}^{TC} C_n \text{ (if } n<=LL) \right] \cdot 100}{N}$$

Equation 7.1

$$\% \text{ OL High} = \frac{\left[\sum_{n=LC}^{TC} C_n \text{ (if } n>UL) \right] \cdot 100}{N}$$

Equation 7.2

LC = Low Cell (input to macro)

CS = Cell Size (input to macro)

TC = LC + CS*30

LL = Low Limit (input with data)

UL = Upper Limit (input with data)

C = Count in cell n

N = Total number of elements

7 Small Sample Statistics

The following example illustrates this process:

Lim	Cnt	≤ =	≥
1	1	1	0
2	2	2	0
3	2	0	0
4	2	0	0
5	2	0	2
Sum	10	3	2

The first column in the table is the actual cell limit. The second column is the count (number of elements in the class). In the third column, the class count is repeated if the cell limit exceeds the minimum. For this example, the lower limit is two, and the upper limit is four.

There are ten elements (the sum of column 2). Three elements are less than or equal to the lower limit, and two are greater than the upper limit. The percents below and above are found by dividing the sums of the last two columns by the number of elements (10), with the results being 30% and 20% respectively. Cells Q24 and Q25 perform this function for the current template. Q26 erases the two columns after the calculations are recorded and converted to numbers.

```
Q23: '/cu94.v94~s47.s75~
Q24: '100*(@sum(s47.s75)/$u$42)~/rff1~~{edit}{calc}{down}~
Q25: '100*(@sum(t47.t75)/$u$42)~/rff1~~{edit}{calc}{down}{down}
Q26: '/res47.t75~/cu95.w95~s47.s75~
```

After erasing the two columns above, cell Q26 copies three new equations into the data table. The equation in cell U95 generates a cumulative value column (column S in the table) based on the values in column R (the cell count). The equation in V95 is copied to column T. This equation produces a 1 when the current line contains the median and a zero otherwise. A cell contains the median if the cumulative contents exceed half of the total values and the cumulative cell below does not exceed half of the total values.

Small Sample Statistics 7

The final column —U— contains the actual median equation (Equation 7.3) in each cell. In column T, the median is multiplied by the contents of the cell. (Remember, the amount in column T will be zero except when the current cell contains the median.) Only one cell will be nonzero, and it will contain the median. The sum of column T will contain that same value. Cell Q27 performs this summation and stores it in the data range. Figure 7.5 shows an excerpt from this table with a sample data set.

$$\text{Median} = BB + \left[CS * \frac{(\text{INT}((N/2)+.5) - CB)}{CC} \right]$$

Equation 7.3

BB = Boundary of cell below
CB = Count below median cell
CC = Count in the median cell
CS = Cell Size
N = Total number of elements

Cell Q28 calculates the Pearsonian coefficient of skewness and stores it below the median in the data output range. This value is found by dividing the difference between the mean and the median by the standard deviation (Equation 7.4).

$$\text{Skew} = \frac{\overline{X} - \text{Mode}}{\sigma}$$

Equation 7.4

Q27: '@sum(u47.u75)~{edit}{calc}{down}
Q28: '(@avg(data)-{up})/@std(data){edit}{calc}{down}

169

7 Small Sample Statistics

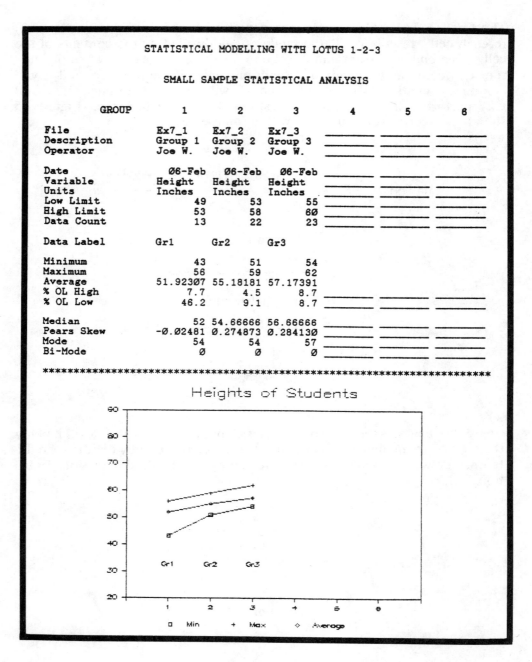

Figure 7.5 Answer to Exercise 7.1

The final calculation is the easiest. Cell Q29 erases the added columns and sorts the original two columns on cell quantities. The mode is the most common value, so, by definition, it will be moved to the top. Cell Q47 will contain the mode because it will always be the top or most frequent value. Q30 moves this value to the top of the data area. Q31 records a 1 in the bimode indicator if the largest value occurs more than once. Other indicators could be substituted here as was previously described. The final lines erase the whole table and return control to the primary menu.

Median/Mode Macro

```
Q29:    '/res47.u75~/dsdq47.r75~pr47.r75~d~g
Q30:    '+q47~{edit}{calc}~
Q31:    '{down}@if(r47=r48,1,0){edit}{calc}~
Q32:    '/req47.r76~
Q33:    '{home}/xmmenu1~
```

Print

The summary table is printed by the following four cells. It is a straightforward print where Q94 clears the existing print settings and then sets the required ranges and executes. The range printed A31.H31 does not change from run to run.

Print Macro

```
Q93:    'Print
Q94:    '/ppcarq86.y89~oouqag
Q95:    'ra1.h31~gq
Q96:    '/xr~
```

The printing of data is accomplished in the Print menu. The cursor is moved to cell A31 and then down nine cells. A window is opened, and a message is displayed, prompting the user to paint the data range to be printed and then restart the macro. The data range is then printed. The following cells are the print portions of the menu and the cells containing the user message.

7 Small Sample Statistics

Screen Labels

```
J49:  '{goto}a31~
J50:  '{down}{down}{down}{down}{down}
J51:  '{down}{down}{down}{down}/wwu/wwh{window}
J52:  '{goto}q100~{window}
J53:  '/ppcar.{?}~
J54:  'gpq~/wwc
J55:  '/xmmenu1~

Q100: 'Print data range must be set manually.
Q101: 'Use arrows to paint data range.
Q102: 'Press return when range is covered.
```

Graph

The Graph mode execution begins in cell Q2. Existing parameters are reset, and the X range is entered. The next line sets the F range to provide the on-screen data labels. Options Data Format is executed, and the data label range is set. Symbol format is set to neither.

```
Q1:  'Graph
Q2:  '/grgxq78.x78~
Q3:  'fq80.x80~odfb18.h18~aqffnqq
```

The manual scale factors are set in Q4 to the contents of cells R82 and R81. The remaining cells provide the labels for the graph and the axis range. View is executed so the graph will appear on the screen at the end of the macro's execution.

Graph Macro

```
Q4:  'osymur82~lr81~q
Q5:  'laMin~
Q6:  'lbMax~
Q7:  'lcAverage~
Q8:  'tf\c41~
Q9:  'ts\c42~
Q10: 'tx\c43~
```

continued...

...from previous page

Q11: `'ty\c44~qq`
Q12: `'/gab20.h20~`
Q13: `'bb21.h21~`
Q14: `'cb22.h22~v`
Q15: `'q{home}/xmmenu1~`

Menu 5 contains the Graph Store function.

RANGE NAMES

Prior to execution of this template, the range names must be entered. The following commands will fulfill this requirement. The companion disk includes these named ranges and is fully operational:

```
/rncbin~q47.q75~
/rncdata~c46~
/rncgraph~q2~
/rncload~u2~
/rncmenu1~I30.n31~
/rncmenu2~I37.n39~
/rncmenu3~I47.I48~
/rncmenu4~I59.o60~
/rncmenu5~I68.k69~
/rncmmc~Q19~
/rncprint~q94~
/rnczcolumn~x47.x71~
/rnc\0~I24~
/rnc\A~I24~
```

OPERATION

Like most of the templates in this book, this template's operation is menu-driven. When specific information is required, the user will be prompted. Two critical values must be entered to provide a basis for median and mode calculations: the minimum cell value and the cell size.

7 Small Sample Statistics

The first value that must be entered is the minimum cell for the frequency distribution. This value generally should be slightly lower than the lowest data value. All values below the minimum cell value will be included in the lowest cell by the 1-2-3 Data Distribution command used to generate the frequency distribution. The second value that must be entered is the cell size. There will always be 28 cells generated, so this second value entered must be large enough to make the top cell value in the distribution encompass even the largest data value.

Data will be spread evenly over the distribution range. If there is a single value very far from the rest, including it may cause the remaining values to be compressed into too few cells. In this case, the outlying value may be deleted, but not without consideration of the effect on median and mode. Ideally, data should cover about fifteen cells, which can be controlled by proper selection of cell size.

Statistical purists will be careful to select the minimum cell value and cell size to prevent any single data value from falling on a cell boundary.

SUMMARY

- To analyze a data set, you should take size and type of data into consideration and should determine those characteristics before data are taken.

- For analysis of small sample sizes, the **minimum**, **maximum**, **average**, **median**, and **mode** techniques are most commonly used.

- The **median** is a value such that half of the data set is above and half below it.

- The **mode** is the value that occurs most frequently.

- A **bimodal** distribution or data set has more than one data group included in the dependent variables data set.

- 1-2-3 does not automatically calculate median and mode. These values must be handled in a **scratchpad** (or calculation area) in the spreadsheet.

- Mode is found by sorting a range, including cell boundaries and counts, on the basis of count. The top boundary value is the mode after this sort is complete.

- Logical AND statements in an @IF function are performed by nesting @If commands.

7 Small Sample Statistics

EXERCISES

Case Study Three: Heights of Students

For the exercises in this chapter, enter the data from Table 7.1, which follows. Data should be entered into three data files with the names indicated.

Dimension	1	1	1
File Name	EX7__1	Ex7__2	EX7__3
Description	GRP1	GRP2	GRP3
Operator	Joe W.		
Variable Name	Height		
Units	Inches		
Low Limit	49	53	55
High Limit	53	58	60
Count	13	22	23
Graph Title 1	Heights of students		
Title 2	Exercise 7.1		
Floating Label	GRP1	GRP2	GRP3
Data	52	54	56
	54	56	57
	53	57	56
	54	56	57
	56	54	58
	51	53	60
	52	59	59
	54	51	62
	54	56	58
	43	57	57
	49	58	54
	51	57	58
	52	54	56
		53	57
		54	58
		55	57
		55	56
		54	55
		55	54
		56	57
		57	56
		53	58
			59

Table 7.1 Data for Chapter 7 Exercises

Small Sample Statistics **7**

1. Load the SMALL template, and enter the three data files created in Table 7.1 into the first three columns. Use 40 for the cell limit and 1 for the cell size. Generate a graph and print the results, including the graph. Print the data.

2. Unload the second two files.

3. Enter the first file again into column two. Use 39.5 for the cell limit and 1 for the cell size.

4. Enter the first file three more times (columns 3-5) using the following for limits:

Cell Limit	Cell Size
40	2
25	1
56	1

   What difference has varying the cell limits made?

Answers

1. The commands that follow will accomplish the results for Exercise #1.

Command	Description
\<return\>	Response to opening screen
f1EX7_1\<return\>	File Load Ex7__1
40\<return\>1\<return\>	Cell Limit and Size
crqflex7_2\<return\>	Cursor Right, File Load
40\<return\>1\<return\>	
crrqflex7_3\<return\>	Cursor Right Right, File Load
40\<return\>1\<return\>	
gc\<return\>gsex7_1\<return\>	Graph Create, Graph Save
pspdpq	Print Summary Page, Data Page Quit

7   Small Sample Statistics

Exit 1-2-3 without moving the paper in the printer. Add the graph just saved by performing the following. Note that the graph size must be set manually.

<p>rintgraph	Enter Lotus Printgraph
IEX7_1<SP><return>	Select Graph
SISM	Settings Image Size Manual
TO<return>	Top Printed with no spaces
L 1.50<return>	Left Margin 1.5 inches
W 6.0<return>	Width 6 inches
H 4.0<return>	Height 4 inches
GPQ	Go Page Quit

2. Files are unloaded through the File Unload command. The cursor determines which file will be removed:

crrq	Cursor to column 3
fu	File Unload
crq	Cursor to column 2
fu	File Unload

3. The cursor should still be in column 2; if not, move it there. Enter the following:

flex7_1	File Load Ex7_1
39.5<return>1<return>	

Use the Cursor Down command to bring the lower half of the data area into view. Note that the minimum, maximum, and average have not changed (they are calculated directly from the raw data). The median has increased to 52.5 since the calculation is based on a cell boundary increased by 0.5. The skew is more negative since the shift in median increased the distance between median and mean. The mode has also shifted, as was expected. Return the cursor to home.

Small Sample Statistics 7

4. Using the same method, enter the file three more times with the different parameters for cell limits:

   crrqflex7_1<return>40<return>2<return>
   crrrqflex7_1<return>25<return>1<return>
   crrrrqflex7_1<return>56<return>1<return>

   Move the cursor down to display the lower portion of the data area In the third column, the bimode indicator is positive. This result was caused by redistributing the data into fewer cells, which also caused column two to have the same count. In the fourth cell, a data error has occurred, causing an apparent reduction in high values. This reduction is due to the low starting cell selected. There are always 28 cells, and the operator must select cell limits to include all data points. In the fifth column, the median, skew, and mode have all increased because the low values were all shifted up into the minimum cell, 56. Again, the operator must use caution in selecting ranges. In a later chapter, a template will be created allowing the frequency distribution to be displayed. Users wanting increased protection can add a line to the current display, which will print the total count from the cumulative column as a double check.

TEMPLATE LISTING

The following listing was transferred to the word processor from a working copy of the template. To enter this template, use this listing, not the segments within this chapter. Before execution, the range names previously described must be entered.

```
A1: '
B2: '          SMALL SAMPLE STATISTICAL ANALYSIS
B5: 'GROUP
C5: ^1
D5: ^2
E5: ^3
F5: ^4
G5: ^5
H5: ^6
```

continued...

7 Small Sample Statistics

...from previous page

```
A7:  'File
A8:  'Description
A9:  'Operator
A11: 'Date
A12: 'Variable
A13: 'Units
A14: 'Low Limit
A15: 'High Limit
A16: 'Data Count
A18: 'Data Label
A20: 'Minimum
A21: 'Maximum
A22: 'Average
A23: '% OL High
A24: '% OL Low
A26: 'Median
A27: 'Pears Skew
A28: 'Mode
A29: 'Bi-Mode

C7:-H9:   '_____
C11:-H12: (D2) '_____
C12:-H16: '_____
C23:-H24: '_____
C26:-H29: '_____
A31:-H31: '********

A33: 'File
A36: 'Variable
A41: 'Graph
A46: 'Data

B32: 'Dmnsion
B33: 'Name
B34: 'Descrip
B35: 'Operator
B36: 'Name
B37: 'Units
B38: 'Lo Limit
B39: 'Hi Limit
```

continued...

Small Sample Statistics 7

...from previous page

```
B40:  'Count
B41:  'Title 1
B42:  'Title 2
B43:  'X-Title
B44:  'Y-Title
B45:  'Flt Label
B46:  'Start

I3:   '         ******************************************************
I4:   '         ******************************************************
I5:   '         ***                                                ***
I6:   '         ***                                                ***
I7:   '         ***      STATISTICAL MODELING WITH LOTUS 1-2-3     ***
I8:   '         ***              COMPANION DISKETTE                ***
I9:   '         ***                                                ***
I10:  '         ***                                                ***
I11:  '         ***          A STATISTICAL SOFTWARE PACKAGE        ***
I12:  '         ***                                                ***
I13:  '         ***                                                ***
I14:  '         ***                                                ***
I15:  '         ***              SMALL SAMPLE TEMPLATE             ***
I16:  '         ***                                                ***
I17:  '         ***                                                ***
I18:  '         ******************************************************
I19:  '         ******************************************************

R39:  'Median/Mode Scratchpad - Current column
R40:  'Minimum
T40:  'Lo Lim
R41:  'Maximum
T41:  'Hi Lim
R42:  'Low Cell
T42:  'Count
R43:  'cls size
T43:  'Median
Q46:  "Bin
R46:  "Count
X46:  'column
```

continued...

7 Small Sample Statistics

...from previous page

```
X47:-X49:  '_____
X51: (D2)  '_____
X52:-X56:  '_____
X63:-X64:  '_____
X66:-X69:  '_____
X71:  '*********

Q86:-X86:  '*********
Q81:  'Low
Q82:  'High
Q88:  '              STATISTICAL MODELLING WITH LOTUS 1-2-3
Q100: 'Print data range must be set manually.
Q101: 'Use arrows to paint data range.
Q102: 'Press return when range is covered.
```

Screen Labels

MACRO LISTINGS

```
I22:  'MACROs
I23:  'start
I24:  '/wtc{home}{goto}i1~
I25:  '/xlPress ENTER to continue ~a1~
I26:  '{home}{goto}c7~/wtb
I27:  '/xmmenu1~
```

Autostart Macro

```
I29:  'MENU1
I30:  'File
I31:  'Load   Unload
I32:  '/xmmenu2~
J30:  'Print
J31:  'Table   Data   Both
J32:  '/xmmenu3~
K30:  'Cursor
K31:  'Move the cursor
```

continued...

...from previous page

```
K32:  '/xmmenu4~
L30:  'Graph
L31:  'Create   Save   Quit
L32:  '/xmmenu5~
M30:  'Recalc
M31:  'Recalculate spreadsheet after data or bin adjustments
M32:  '{pgdn}{pgdn}{up}{up}{up}
M33:  '/xgu6~
N30:  'Quit
N31:  'Return to main menu
N32:  '/fdb:\~
N33:  '/frmenu~
```
                        Menu 1 Listing


```
I36:  'Menu2
I37:  'Load
I38:  'Load a file in the current column
I39:  '/xgload~
J37:  'Unload
J38:  'Remove file in the current column
J39:  '/re{pgdn}{pgdn}{pgdn}{end}{down}~
J40:  '/czcolumn~~
J41:  '/xmmenu1~
K37:  'Directory
K38:  'Change the data directory
K39:  '/fd{?}~
K40:  '/xmmenu1~
L37:  'Stop
L38:  'Stop execution to adjust parameters
L39:  '/wtc{home}
L40:  '/xq
M37:  'Quit
M38:  'Return To Main Menu
M39:  '/xmmenu1~
```
                        Menu 2 Listing

continued...

7 Small Sample Statistics

...from previous page

```
I45:  'Menu 3
I46:  'Summary
I47:  'Print a summary report
I48:  '/xcprint~
I49:  '/xmmenu1~
J46:  'Data
J47:  'Print summary report with data
J48:  '/xcprint~
J49:  '{goto}a31~
J50:  '{down}{down}{down}{down}{down}
J51:  '{down}{down}{down}{down}/wwu/wwh{window}
J52:  '{goto}q100~{window}
J53:  '/ppcar.{?}~
J54:  'gpq~/wwc
J55:  '/xmmenu1~
K46:  'Page
K47:  'Page feed printer
K48:  '/pppq~
K49:  '/xmmenu1~
L46:  'Quit
L47:  'Return to main menu
L48:  '/xmmenu1~
```

Menu 3 Listing

```
I58:  'Menu 4
I59:  'Left
I60:  'Cursor left
I61:  '{left}/xmmenu4~
J59:  'Right
J60:  'Cursor right
J61:  '{right}/xmmenu4~
K59:  'Up
K60:  'Cursor up
K61:  '{up}/xmmenu4~
L59:  'Down
L60:  'Cursor down
L61:  '{down}/xmmenu4~
M59:  'Go To
M60:  'Go to a specific cell location
```

continued...

...from previous page

```
M61: '{goto}{?}~
M62: '/xmmenu1~
N59: 'Home
N60: 'Home cursor
N61: '/wtc{home}{goto}c7~/wtb~
N62: '/xmmenu1~
O59: 'Quit
O60: 'Return to Menu1
O61: '/xmmenu1~
```
Menu 5 Listing

```
I66: 'Menu 5
I67: 'Create
I68: 'Create a graph
I69: '/xggraph~
J67: 'Save
J68: 'Store a graph
J69: '/gs{?}~q
J70: '/xmmenu1~
K67: 'Quit
K68: 'Return to main menu
K69: '/xmmenu1~


Q1:  'Graph
Q2:  '/grgxq78.x78~
Q3:  'fq80.x80~odfb18.h18~aqffnqq
Q4:  'osymur82~lr81~q
Q5:  'laMin~
Q6:  'lbMax~
Q7:  'lcAverage~
Q8:  'tf\c41~
Q9:  'ts\c42~
Q10: 'tx\c43~
Q11: 'ty\c44~qq
Q12: '/gab20.h20~
Q13: 'bb21.h21~
Q14: 'cb22.h22~v
Q15: 'q{home}/xmmenu1~
```
Graph Macro

continued...

7 Small Sample Statistics

...from previous page

```
Q18: 'Median/Mode calculations
Q19: '/xnEnter the Top of the lowest cell ~s42~
Q20: '/xnEnter the cell size ~s43~
Q21: '/dfbin~s42~s43~~
Q22: '/dddata~bin~
Q23: '/cu94.v94~s47.s75~
Q24: '100*(@sum(s47.s75)/$u$42)~/rff1~~{edit}{calc}{down}~
Q25: '100*(@sum(t47.t75)/$u$42)~/rff1~~{edit}{calc}{down}{down}
Q26: '/res47.t75~/cu95.w95~s47.s75~
Q27: '@sum(u47.u75)~{edit}{calc}{down}
Q28: '(@avg(data)-{up})/@std(data){edit}{calc}{down}
Q29: '/res47.u75~/dsdq47.r75~pr47.r75~d~g
Q30: '+q47~{edit}{calc}~
Q31: '{down}@if(r47=r48,1,0){edit}{calc}~
Q32: '/req47.r76~
Q33: '{home}/xmmenu1~
```
                    Median/Mode Macro

```
U1:  'Load
U2:  '{down}{down}{down}{down}
U3:  '@today~/rfd2~{pgup}
U4:  '{pgdn}{pgdn}{up}{up}{up}
U5:  '/fcce{?}~
U6:  '{down}
U7:  '/c{down}{down}~
U8:  '{pgup}{pgup}{down}{down}~
U9:  '{down}{down}{down}/c{down}{down}{down}{down}~
U10: '{pgup}{Pgup}{down}
U11: '{down}{down}{down}~
U12: '{DOWN}{DOWN}/c{down}~u40~
U13: '{DOWN}{DOWN}{DOWN}{down}
U14: '{DOWN}{DOWN}{DOWN}
U15: '/c~{pgup}{pgup}{down}~
U16: '{down}/rnddata~/rncdata~.{end}{down}~
U17: '{pgup}{up}{up}{up}{up}{up}{up}
U18: '{up}{up}{up}{up}{up}
U19: '{up}{up}{up}{up}{up}
U20: '@count(data)~{edit}{calc}~/c~u42~
U21: '{down}{down}{down}{down}
```

continued...

...from previous page

```
U22:  '@min(data)~{edit}{calc}~/c~s40~{down}
U23:  '@max(data)~{edit}{calc}~/c~s41~{down}
U24:  '@avg(data)~{edit}{calc}~{down}
U25:  '/xgmmc~
```
                    Load Macro


```
Q93:  'Print
Q94:  '/ppcarq86.y89~oouqag
Q95:  'ra1.h31~gq
Q96:  '/xr~
```
                    Print Macro


```
Q77:  'Graph limit constants
R78:  1
S78:  2
T78:  3
U78:  4
V78:  5
W78:  6
Q79:  1.2*@MAX(C21..H21)
R79:  (Q79)                         Copy From:R79    To:S79.X79
Q80:  @IF($C$20,1.5*$C$20,0.7*$C$20)
R80:  (Q80)                         Copy From:R80    To:S80.X80
R81:  @IF(@MIN(C20..H20),2*@MIN(C20..H20),0.5*@MIN(C20..H20))
R82:  @IF(@MAX(C21..H21)0,1.4*@MAX(C21..H21),0.7*@MAX(C21..H21))
U94:  @IF(S94U$40,T94,0)
V94:  @IF(S94=$U$41,T94,0)
U95:  (U94+T95)
V95:  @IF(U95=$U$42/2,@IF(U94U$42/2,1,0),0)
W95:  @IF(V950,(S95-$S$43)+($S$43*(@INT(($U$42/2)+0.5)-U94)/T95),0)
```
                Scratchpad and calculations

CHAPTER 8

DESCRIPTIVE STATISTICS

8 Descriptive Statistics

The methods presented in this chapter are the most commonly used and understood of all statistical methods available. Almost everyone knows what the mean and standard deviation are and how to use them. These calculations are so common that Lotus has included functions to automate both of them. The standard functions provide the mean of the population and its standard deviation. A worksheet is included in this template to calculate the unbiased or n-1 weighted mean and standard deviation.

The Descriptive Statistics template demonstrated here will integrate these functions into a summary report, which includes the following:

- Means (sample and population)
- Standard deviations (sample and population)
- Variances
- Kurtosis and Skewness (third and fourth moments)
- Frequency distribution table
- Histogram of points in the range

The calculations for means and moments of the distributions, which will be explained in this chapter, are performed on the actual data values, not on the cell boundaries. The frequency distribution table includes a histogram-type output that can be added to any summary report to form a single-page output report. Optionally, a bar graph can be appended to the summary report for a graphic presentation without the data table.

Figure 8.1a is a report generated with the frequency distribution table, and 8.1b is the same report with a histogram. It is obvious from the report that the cell size is incorrect. Every other cell in the data range is zero. The resolution of the plot exceeds the resolution of the measurement. The abilities to correct granularity errors (change starting bin and bin size) and replot the data are included. This information is stored in a spreadsheet called DESCRIPT.

Descriptive Statistics 8

```
            STATISTICAL MODELLING WITH LOTUS 1-2-3

                  DESCRIPTIVE STATISTICAL ANALYSIS

          Variable: Height    inches    Quantity:       38
                File: Examp8_1

                              N Weight              N-1 Weight
              Mean:  . . . . .69.9210526  . . . . .69.9210526
           Std Dev:  . . . . .2.35507938  . . . . .2.38669259
          Variance:  . . . . .5.54639889  . . . . .5.69630156
           Minimum:  . . . . .        65  . . . . . . . . . .
           Maximum:  . . . . .        75  . . . . . . . . . .
             Range:  . . . . .        10  . . . . . . . . . .
          Skewness:  . . . . .0.07038441  . . . . . . . . . .
          Kurtosis:  . . . . .-0.5173588  . . . . . . . . . .

             Descr: Example 8-1, Heights of individuals
          Operator: Will                    Date:  04-Jul-76
*****************************************************************
   Cell                         10%  20%  30%  40%  50%  60%  70%  80%
  Limit  Count     %    Cum      |    |    |    |    |    |    |    |
     64     0    0.0   0.0    |.
   64.5     0    0.0   0.0    |.
     65     1    2.6   2.6    |+
   65.5     0    0.0   2.6    |.
     66     2    5.3   7.9    |++
   66.5     0    0.0   7.9    |.
     67     3    7.9  15.8    |+++
   67.5     0    0.0  15.8    |.
     68     5   13.2  28.9    |++++++
   68.5     0    0.0  28.9    |.
     69     5   13.2  42.1    |++++++
   69.5     0    0.0  42.1    |.
     70     7   18.4  60.5    |+++++++++
   70.5     0    0.0  60.5    |.
     71     6   15.8  76.3    |+++++++
   71.5     0    0.0  76.3    |.
     72     3    7.9  84.2    |+++
   72.5     0    0.0  84.2    |.
     73     3    7.9  92.1    |+++
   73.5     0    0.0  92.1    |.
     74     2    5.3  97.4    |++
   74.5     0    0.0  97.4    |.
     75     1    2.6 100.0    |+
   75.5     0    0.0 100.0    |.
     76     0    0.0 100.0    |.
   76.5     0    0.0 100.0    |.
     77     0    0.0 100.0    |.
   77.5     0    0.0 100.0    |.
     78     0    0.0 100.0    |.
   78.5     0    0.0 100.0    |.
     79     0    0.0 100.0    |.
Low cell contains points below distribution
Pts above High cell       0
Total Points in dist     38
```

Figure 8.1a Sample Output from Descriptive Statistics Template

8 Descriptive Statistics

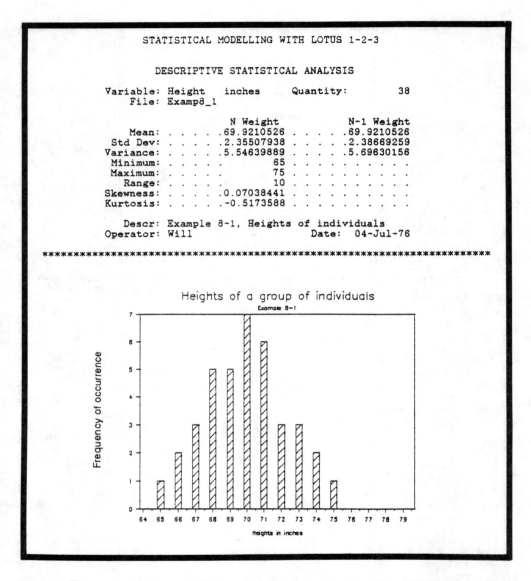

Figure 8.1b Sample Output from Descriptive Statistics Template

When this program is run, the user is asked to input the low cell and cell size. The low cell should typically be slightly lower than the lowest data value. The cell size controls the range of values that will be spread over 28 cells. The user can replot any histogram with different cell limits or sizes, including boundaries away from actual values.

MENUS

Autostart

The two menu levels for this spreadsheet are shown in Figure 8.2. The autostart menu provides six alternatives, one of which is to Quit and return to the main system menu. The remaining five options are all submenus: File, Print, Graph, Cursor, and Adjust.

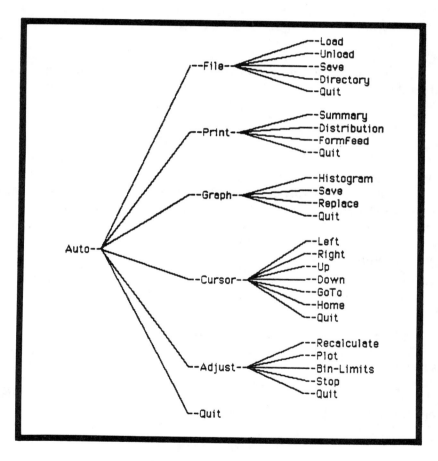

Figure 8.2 Descriptive Statistics Menu

File

The **File** menu allows the user to **Load** a file from disk into the spreadsheet for analysis. This file can also be (U)nloaded. Unload is a necessary operation since the /FC (File Combine) operation is used to load files. Remember, FileCombine intermixes the new file with whatever exists in the spreadsheet. Unload should be performed prior to File Load. **Save** allows the current data set to be stored to a disk file, which is useful because data files can be created or changed while in the DESCRIPT template and then saved to disk. The **Directory** command allows the user to select the data file location. If you have a hard disk, you are better off using a data disk in drive A or B and storing programs on the hard disk. With a two-floppy system, using drive B for programs and drive A for data disks is also recommended. Doing so requires removing the Lotus disk from drive A after boot-up but has the advantage of keeping data on a separate disk from the programs or Lotus itself.

Print

The **Print** menu allows a summary report to be printed to LPT1. In some instances, this output is sufficient. If Distribution is requested, the frequency distribution table (see Figure 8.1a) is appended to the summary report. This report is designed to take exactly one page. If **FormFeed** is selected, a Top of Form command is sent to the printer. Do not make this selection if you want to append a graph to the summary report.

Graph

The **Graph** menu contains options to generate the histogram report and save it to disk. If this option is selected, a graph can be created and displayed on the screen. The **Save** and **Replace** options allow this picture to be stored. Use Save the first time a file name is used and Replace if the file already exists. If an error occurs while running this macro, press <Esc> and clear the screen; then, restart by pressing <Alt/A>.

Cursor

The **Cursor** commands allow the user to scroll **Up**, **Down**, **Right**, or **Left** one page at a time to view portions of the spreadsheet. The Goto command allows the user to move to a specific location on the screen. Moving to the range named Dist allows the user to look at the frequency distribution table on-screen prior to printing. The **Goto** command cancels any titles that limit the cursor's mobility in the display area. Execute a **Home** command after any Goto or to return the cursor to Home from anywhere in the program, and the titles will automatically reset.

Adjust

The **Adjust** menu allows recalculation of the entire spreadsheet after data changes have been made. There can be many reasons for making data changes. Starting from scratch, raw data can be entered, values can be changed, graph labels can be changed, etc. After any of these changes, restart the macro by pressing <Alt/A>, and then execute **Recalculate**. **Plot** recalculates the frequency distribution table with new **bin limits**. Remember, bin limits allow manual setting of the low cell and bin size for the histogram and frequency distribution table. Familiarity with the operation of this function is extremely useful. The **Stop** function halts the macro after clearing the titles and executing a Home. This command allows the user the freedom to move anywhere on the spreadsheet and make changes. It is especially handy for data entry and corrections.

LAYOUT OF THE TEMPLATE

Figure 8.3 illustrates the layout of the Descriptive Statistics template. It is four screens wide (32 columns). In several areas, column width adjustments have been made. Primarily, these adjustments have been made in the frequency distribution table, starting in cell AA3, where it was necessary to make the columns readable and limit them to one page.

8 Descriptive Statistics

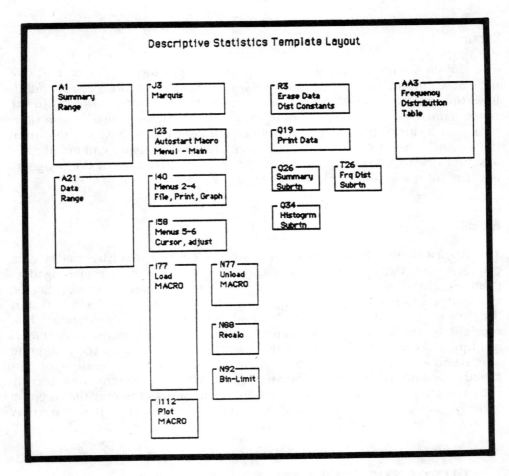

Figure 8.3 Descriptive Statistics Template Block Diagram

METHODS AND APPLICATIONS

Purists in the field of statistics insist that the standard deviation of a number should not be used unless 32 points or more are used in its calculation. Histograms should have a minimum of 16 cells, and cell limits should be selected so that no point falls on a boundary. In the example of Figure 8.1, where resolution of the measurement was one inch, proper cell boundaries would be half inches around the range of 64" to 80", with a 1" cell size. With measurements to the nearest inch and boundaries on the half inch, no cell would fall on a cell edge.

Descriptive Statistics 8

These guidelines are excellent and should be followed when possible, but not without some understanding of why they are suggested. For small quantities, the standard deviation is sensitive to change. In pulling samples from a normal distribution, probability that a single value will be close to the mean is high, and deviation will be small. If one of the values in a small sample is near the three sigma point, the deviation calculated will be artificially high. When 30 samples are used in the calculation, a value near the three sigma point has little effect.

The last chapter showed why for these small data sets the Min-Max-Average analysis is effective as an indicator of how well the data points are centered. If the mean (average) is centered between the minimum and maximum, it is probably not skewed. Comparison of the mean to the median further indicates whether or not the mean is skewed. For larger samples, the third moment — or skewness — indicates a shift in distribution.

Mean and Standard Deviation

A wide variety of methods is used to make the various outputs seen in Figure 8.1a. Minimum, Maximum, Average, Standard Deviation, Variance, and Count are made using standard Lotus @ functions on the data range. Range is computed from the Minimum and Maximum. The remaining values are calculated using the equations in Figure 8.4.

8 Descriptive Statistics

8.1 $\quad \sigma = \sqrt{\frac{\Sigma(X-\bar{X})^2}{n-1}}$

8.2 $\quad S = \frac{m_3}{\sigma^3} = \frac{\Sigma(X-\bar{X})^3}{n} \div \sigma^3$

8.3 $\quad K = \frac{m_4}{\sigma^4} = \frac{\Sigma(X-\bar{X})^4}{n} \div \sigma^4 - 3$

K = Kurtosis
m_n = nth statistical moment
n = number of samples
S = Skew
σ = Standard deviation
X = Sample value
$\bar{X}$ = Sample average

Figure 8.4 Descriptive Statistics Equations

Descriptive Statistics 8

Unbiased Estimator

Equation 8.1 (in Figure 8.4) is the **unbiased** standard deviation. To compute this equation, generate a column of values containing the difference between each value in the data range and the mean value. This step can be accomplished in one pass, as the mean has already been determined using @Avg. The sum of this column is divided by the square root of the number of elements in the range minus one. The result is the unbiased – or n-1 weighted – standard deviation. The difference between the biased and unbiased values is the ratio of the denominators shown in Figure 8.5.

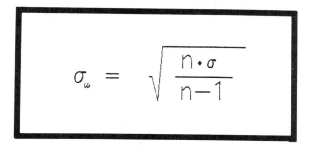

Figure 8.5 Unbiasing Standard Deviations

The three graphs in Figure 8.6 are plots of the probability density function for three sets of data with different sample sizes but with the same mean and standard deviation. In the first graph, the five sample sizes show nearly a 10% difference between biased and unbiased deviation. When the PDF is plotted, the distribution is flatter and wider for the unbiased function. In the second pair of curves, with a 15-point sample size, the difference is much less; the ratio of deviations is down to 4.5%. In the plot for the distribution with 35 points, the deviations differ by only 1.6%. The two plots are nearly identical, which is another reason to work with 30 or more points.

8 Descriptive Statistics

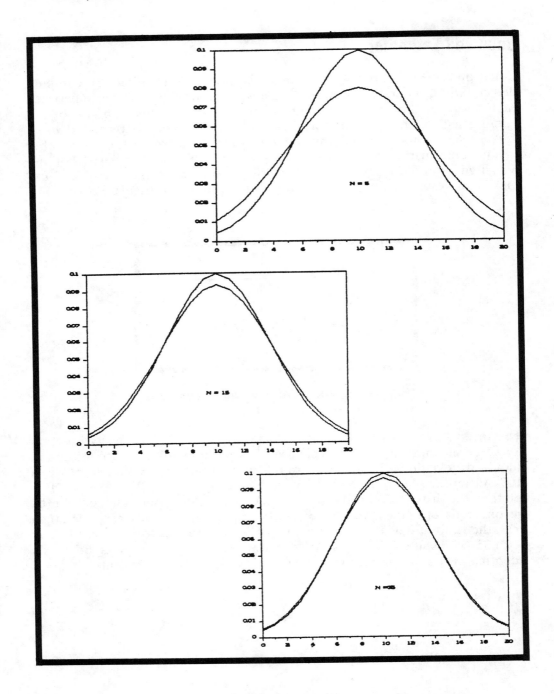

Figure 8.6 Biased vs. Unbiased Deviations — Different Sample Sizes

Moments

Equations 8.2 and 8.3 in Figure 8.4 demonstrate **skewness** and **kurtosis**. The summation term in each equation is the **moment** in a manner similar to moments of inertia in physics. In mechanics (physics), the moments are weighted sums calculated from forces and the distance over which they are applied. In statistics, the summation — or moment — is weighted by its distance from the mean value and sample size. This equation should look familiar, because the standard deviation — or second moment — is m_2. Concurrent with the calculation of the unbiased standard deviation above, the skewness and kurtosis are calculated. Two additional columns are created with the difference between data elements and the mean data value raised to the third and fourth powers included. Summing these columns provides the numerators for the moment terms in the two equations. As was explained earlier, this summation can be accomplished in a single pass because Lotus provides the mean of the data elements with a single command. The @count function provides the denominator, and sigma is the standard deviation already calculated.

Skewness

Skewness is a measure of the tilt of the data set. Like the skewness measurement in the last chapter, it indicates whether or not there is bias on the data set or if an outlying value is distorting it. In a symmetrical distribution, skewness will be zero when this equation is used. If more values occur to the right of the mean, the skew is positive, and if more occur to the left, the skew is negative. In practice, a number of 1.5 positive or negative is a fairly large skew for a normal distribution. (See Equation 8.2.)

Kurtosis

Kurtosis is a measure of the peakedness of the distribution. A good visual example of this concept is in the three sets of graphs in the last section. The difference between the biased and unbiased standard deviations is one of kurtosis. The biased deviation is more peaked (less spread). Since a normal, or Gaussian, curve has a kurtosis of three, the third equation has a -3 built in. Using this equation, then, a normal curve has a kurtosis of zero. Negative values indicate curves that are flatter than normal, which is referred to as **platykurtic**. Positive values indicate peaked, or **leptokurtic**, distributions. **Mesokurtic** distributions are those near normal. (See Equation 8.3.)

8 Descriptive Statistics

Details of how these methods are applied will be furnished in the section of this chapter that describes macros. A close look at the layout sheet for this template reveals much open space, which is no accident. The templates for the next two chapters will be added into this spreadsheet to make use of the calculations already made.

Histograms

A frequency distribution table is generated at the end of the Load macro. This chart is generated in a range by itself, starting with cell AA3. The first column is the cell-limit indicator. This indicator is generated using /DF (Data Fill). Its cell size is determined by the user input, as is the starting cell.

Histograms spread over too few cells tend to concentrate their data into one or two cells. Histograms spread over too many cells tend to have a thin data distribution with most cells containing only a few values. If normal (Gaussian) data are spread over 12 to 20 cells, considerable information can be gained about the shape of the data distribution. If cell boundaries are selected so that no value falls on the edge of a cell, there is never a question as to whether a point belongs in the cell above or below it. If values are allowed to fall on the edge of the cell, a slight bias is entered into cellular calculations — like the median in Chapter 7 — but such problems are small, especially as sample size increases.

The second column in the frequency distribution table is the frequency of occurrence or **count**. Each bin contains all the cells below it except those in a lower bin. The bottom bin contains all the points below it, even if they are outliers, which is the way the /DD (Data Distribution) command works. The third column is the count, normalized to the total number of elements and multiplied by 100 to display as a %. The fourth column is a cumulative percent. To the right of this table is a wide column in +/- format, which is in itself a histogram of the data in the table. Each + equals 2%, with the horizontal scale marked to 80%.

In addition to this chart, a bar graph is available. Appending this graph to an output report requires the user to generate the graph from the graph submenu and exit 1-2-3 to use Printgraph. In Figure 8.1b, where this graph was printed, the graph range was set manually to make it slightly smaller than a normal half-page graph. The left margin is increased, while graph height and width are reduced. These adjustments give the graph a better overall appearance.

Descriptive Statistics 8

OUTPUTS

The outputs have already been discussed but will be summarized here for clarity. The summary report is slightly less than half a printed page and is identical to the screen display when in the home position. It includes all the basic statistical calculations.

A frequency distribution table can be appended to this report or printed separately. It includes 30 cells with all the values spread according to user-selected limits. The information in this table can be printed to a graph with a single command.

COMMANDS USED

Following is the list of commands used in the template created in this chapter:

Command	Description
{edit}{calc}	Converts cell contents to value
@std()	Computes standard deviation
@now	Prints today's date (CPU clock)
@var()	Computes variance
/fcce{?}~	Merges file into spreadsheet
/fdb:\~	Changes default directory to B:
/fx{?}~	eXtracts a file or segment
/gs{?}~rq	Saves a graph
/rncdata~	Creates a range named data
/rnddata~	Deletes a range named data
/xc	Macro gosub
/xg	Macro Goto
/xl	Macro label input
/xn	Macro number input
/xr	Macro return from subroutine
/xm	Macro menu
/wtb	Sets both titles
/wtc	Clears titles

8 Descriptive Statistics

MACRO DESCRIPTIONS

The sections that follow give details on how selected macros operate and what support statements they require in the spreadsheet. Not every section will be described in detail, nor will every statement in the covered ranges be described. Only functions that have not been described in this book or that are unique or complicated will be described in depth. A complete listing of the template follows the chapter.

The first section of the template listing, "Spreadsheet Labels and Marquis Listing," is an accumulation of label statements from around the spreadsheet. Primarily, those statements are found in blocks A1, A21, J3, R3, Q19, and part of AA3 on the layout in Figure 8.3. Cell F18, the @now, is the only nonlabel statement in the range.

Autostart

The autostart macro homes the cursor and then moves it to I1 to display the marquis. The macro label (/xl) command puts "Press RETURN to continue" in the command line and halts execution. When <return> is pressed, a blank is stored in cell A1. The cursor is then sent to cell C8 (via home) by line I26. Titles are set to restrict the cursor (by I27), and menu 1 is called.

```
I23:    'start
I24:    '/wtc{home}{goto}i1~{goto}k15~/cI32~
I25:    '/xl Press RETURN to continue~a1~
I26:    '{home}{goto}c8~
I27:    '/wtb~
I28:    '/xmmenu1~
```

Menu Functions

The directory section of the file menu allows the user to change the default drive. Line L43 executes the /FD (File Directory) command and prompts the user for the name of the directory to be installed.

```
L41:    'Directory
L42:    'Change the directory
L43:    '/fd{?}~
L44:    '/xmmenu1~
```

Line I55 calls a subroutine — "histogr" — that takes the data from the frequency distribution table and makes a histogram graph. The results are displayed on the screen. This file can be saved to disk as a PIC file for printing with Lotus Printgraph. Line J55 is the Save option, and K55 is the Replace option. When saving a graph that already exists on disk, use Replace.

```
I52:   'Menu4 - Graph
I53:   'Histogram
I54:   'Create a Histogram graph
I55:   '/xchistogr~
I56:   '/xmmenu4~
J53:   'Save
J54:   'Create a PIC file from the current graph
J55:   '/gs{?}~q
J56:   '/xmmenu4~
K53:   'Replace
K54:   'Save an update to an existing PIC file
K55:   '/gs{?}~rq
K56:   '/xmmenu4~
L53:   'Quit
L54:   'Return to main menu
L55:   '/xmmenu1~
```

Lines I61 and J61 cause the cursor to move five cells left or right respectively. K61 and L61 move up or down one page each. Remember that the distance the cursor moves on a page command differs if worksheet titles are set. In the Goto option at line M61, the titles are cleared prior to executing the Goto. Clearing the titles frees the full screen for viewing portions of the spreadsheet, which is especially useful for looking at the frequency distribution table. On a Goto command, several options are available: enter the cell and <return>; enter the name and <return>; or press F3 (the name key) and select from the menu. The Home command reinstates the worksheet titles.

8 Descriptive Statistics

```
I58:    'Menu5 - Cursor
I59:    'Left
I60:    'Move cursor left five columns
I61:    '{left}{left}{left}{left}{left}
I62:    '/xmmenu5~
J59:    'Right
J60:    'Move cursor right five columns
J61:    '{right}{right}{right}{right}{right}
J62:    '/xmmenu5~
K59:    'Up
K60:    'Move cursor up one page
K61:    '{pgup}
K62:    '/xmmenu5~
L59:    'Down
L60:    'Move cursor down one page
L61:    '{pgdn}
L62:    '/xmmenu5~
M59:    'GoTo
M60:    'Go to a specific cell by name or number
M61:    '/wtc
M62:    '{goto}{?}~
M63:    '/xmmenu5~
N59:    'Home
N60:    'Home cursor
N61:    '{home}
N62:    '{goto}c8~/wtb
N63:    '/xmmenu5~
O59:    'Quit
O60:    'return to main menu
O61:    '{home}
O62:    '{goto}c8~/wtb
O63:    '/xmmenu1~
```

The Recalculate option allows the spreadsheet data to be recalculated after a change is made. Macro operation can be restarted with <Alt/A>. Plot recalculates only the frequency distribution table and is used to change bin limits or create a graph with a change in labels. Halt stops the macro and clears the title, which stops execution and leaves the spreadsheet loaded. Don't confuse Halt with Quit in the primary menu. Quit reloads the Main Menu spreadsheet.

Descriptive Statistics 8

```
I65:    'Menu6 - Adjust
I66:    'Recalculate
I67:    'Recalculate spreadsheet after halt or data change
I68:    '/xg\r~
J66:    'Plot
J67:    'Replot Histogram after limits change
J68:    '/xgplot~
K66:    'Bin-limits
K68:    '/xgbinlim~
L66:    'Stop
L67:    'Halt the MACRO to make changes
L68:    '/wtc{home}
L69:    '/xq
M66:    'Quit
M67:    'Return to primary menu
M68:    '/xmmenu1~
```

Load

The first group of commands in the Load macro fill in the blanks in the summary table. File names and labels are entered, the data range is identified and named (I85), and the statistical functions are entered (I86-I91).

```
I77:    'Load MACRO
I78:    '{goto}c22~
I79:    '/fcce{?}~
I80:    '{down}/c~c5~
I81:    '{down}/c.{down}~c17~
I82:    '{down}{down}/c~c4~
I83:    '{down}/c~d4~
I84:    '{goto}c36~
I85:    '/rnddata~/rncdata~.{end}{down}~
I86:    '{goto}f4~@count(data)~
I87:    '{goto}d8~@avg(data)~
I88:    '{down}@std(data)~
I89:    '{down}@Var(data)~
I90:    '{down}@min(data)~
I91:    '{down}@max(data)~
I92:    '{down}+d12-d11~
```

Cells I93 and I94 are user inputs. In I93 (a numeric input), the user is prompted for the low cell value, which is stored in cell S14. In the same manner, I94 stores the cell size in cell S15.

207

8 Descriptive Statistics

I93: '/xnenter low cell for data distribution~s14~
I94: '/xnenter cell size~s15~

Cell I95 moves the cursor to D36 — the cell immediately to the right of the first data value (in C36) — and loads D36 with the difference between C36 (the first data value) and $D$8 (the average of values in the data column) squared. Remember that the dollar signs make the reference to D8 absolute, so when the cell contents are copied, subsequent cells will also refer to cell D8 for the average value. The next two cells in the macro create additional cells to the right with this difference raised to the third and fourth powers. I98 returns the cursor to D36, where I99 initiates a copy (/C). The three cells just created are copied down the column, creating a table of values. The following sequence,

{left}{end}{down}{right}

moves the cursor to the data range and then down, to the last value, then right, to the point in the d column next to the last value. As a result, the table generated has a row for each data value to enter the range without operator intervention. This four-stroke sequence always copies to a column of cells exactly as long as the data column to its left.

I95: '{goto}d36~(c36-d8)^2~
I96: '{right}(c36-d8)^3~
I97: '{right}(c36-d8)^4~
I98: '{left}{Left}
I99: '/c.{right}{right}~.{left}{end}{down}{right}~

The next two lines sum the three columns, leaving the results in the row above the column. These results are the numerators for the unbiased standard deviation and the two moments. The equations are in Figure 8.4.

I100: '{up}@sum({down}.{end}{down})~
I101: '/c~.{right}{right}~

Line I102 moves the cursor to cell C34 and computes unbiased deviation (the square root of the sum in C35 divided by $F$4-1). F4 is the count of elements in the data range, so the value returned is n-1. I103 converts this cell from an equation to a fixed value, which is copied to the data table (F9). If the conversion-to-value step were omitted, the table could not be erased in cell I106, and considerable slowing of operations would result. The final two lines calculate skew and kurtosis in a similar manner.

```
I102:   '{up}@sqrt({down}/($f$4-1))~
I103:   '{edit}{calc}~/c~f9~
I104:   '{right}(({down}/$f$4)/($d$9)^3)~{edit}{calc}~/c~d14~
I105:   '{right}(({down}/$f$4)/($d$9)^4)-3~{edit}{calc}~/c~d15~
```

The final five lines erase the data table created and move the cursor to place the average and variance in the n-1 weighted column. It is wise to erase the data table because it contains sufficient elements to slow down operation of the spreadsheet.

```
I106:   '{goto}d33~/re.{right}{right}{end}{down}~
I107:   '{home}{goto}f8~
I108:   '@avg(data)~
I109:   '{down}{down}((f9)^2)~{home}
I110:   '/xgplot~
```

The plot routine fills the column from Y3 to Y33 with data, starting with the value stored in S14 earlier (I93) and stepping by the value in S15 (from I94). The range is named Bin, and then a /DD (Data Distribution) command is executed with Data as the data range and Bin as the bin range. The result is the frequency distribution table starting in cell AA3. The two columns generated here provide all the information needed to complete the table.

```
I112:   'Plot-makes distribution and histogram tables
I113:   '/dfy3.y33~s14~s15~~
I114:   '/rncbin~y3.y33~
I115:   '/dddata~bin~
I116:   '/xmmenu1~
```

8 Descriptive Statistics

Unload

This macro erases all data and calculation ranges associated with a data file, which clears the way for a new data file to be entered.

```
N77:    'Unload MACRO
N78:    '/rec4.d5~
N79:    '/rec17.c18~
N80:    '/ref4~/rek15~
N81:    '/rec22.c35~
N82:    '/redata~
N83:    '/cr3.u10~c8~
N84:    '{home}
N85:    '/xmmenu1~
```

Utility Subroutines

Recalculate moves the cursor to the first data element and then transfers control to the load routine at the line following the File Combine.

```
N88:    'Recalculate
N89:    '{goto}c22~
N90:    '/xgi69~
```

BinLim allows the bin minimum and cell width to be changed. A plot should be executed after this command.

```
N92:    'Binlim
N93:    '/xnEnter lower bin limit~s14~
N94:    '/xnEnter bin size~s15~
N95:    '/xmmenu1~
```

Descriptive Statistics 8

Summary prints the summary report.

```
Q26:    'Subroutine Summary
Q27:    '/pplrq19.x22~g
Q28:    'ra1.h20~g
Q29:    '/xr
```

This subroutine creates the histogram graph. X and A ranges are taken from the frequency distribution table, and titles are taken from the data file.

```
Q34:    'Sub-MACRO Histogr
Q35:    '/gxy3.y34~
Q36:    'az3.z34~
Q37:    'tb
Q38:    'otf\c31~
Q39:    'ts\c32~
Q40:    'tx\c33~
Q41:    'ty\c34~q
Q42:    'ncone~vq
Q43:    '/xr
```

The final subroutine sends the frequency distribution table to the line printer.

```
T26:    'Subroutine Histog
T27:    '/ppry1.ad33~g
T28:    'ry36.ad38~g
T29:    'rq19.x19~gpq
T30:    '/xr
```

Calculations for Frequency Distribution

The first two lines that follow are single-cell functions placing totals at the bottom of the chart. The next three are the first cells in the columns in the table. These cells should be entered and then copied 30 lines down.

8 Descriptive Statistics

```
AB37: (+Z35)
AB38: @SUM(Z3..Z34)
AA3:-AA34: (F1) (Z3/$AB$38)*100     (Note: /RFF1 - 1 decimal place)
AB3:-AB34: (F1) (AB2+AA3)     (Note: /RFF1 - 1 decimal place)
AD3:-AD34: (+) (AA3/2)  (Note: /RFF+ - Format +/- width 32)
```

RANGE NAMES

After the spreadsheet information is entered, some final adjustments are needed to make it perform. The following list should be entered as shown. The commands in the first group are range names for the Macro Goto commands; the second group contains column width adjustments.

```
/rncbinlim<return>n93<return>
/rnchistogr<return>q35<return>
/rncload<return>i78<return>
/rncmenu1<return>i35.n36<return>
/rncmenu2<return>i41.m42<return>
/rncmenu3<return>i47.l48<return>
/rncmenu4<return>i53.l54<return>
/rncmenu5<return>i59.o60<return>
/rncmenu6<return>i66.m67<return>
/rncplot<return>i113<return>
/rncsummary<return>q27<return>
/rnctable<return>y1<return
/rncunload<return>n78<return>
/rnc\0<return>i24<return>
/rnc\a<return>i26<return>
/rnc\r<return>n89<return>
```

For each of the following columns, perform the following steps:

- Move the cursor to the first row in the column.
- Execute /wcs (enter the number) <return>.

Descriptive Statistics 8

Column	Width
A	10
D	11
F	11
H	3
Z	5
AA	6
AB	7
AC	1
AD	44

SUMMARY

- Descriptive statistics — including mean and standard deviation — are the most commonly used statistical methods.

- 32 cells is a good guideline for when to use descriptive techniques rather than small sample techniques.

- With 32 elements, the difference between biased and unbiased deviations is 1.6 %.

- Histograms are most effective with 15-20 cells and 30 or more elements.

- Statistical moments are similar to mechanical moments in physics.

- The first statistical moment is always zero by definition — the sum of the differences between individual elements and the mean element is zero.

- The second statistical moment is the standard deviation squared.

- The third moment divided by the cube of standard deviation is **skewness** — a measure of tilt in a data set.

- The fourth moment divided by the deviation raised to the fourth power is **kurtosis** — a measure of peakedness in a data set.

continued...

8 Descriptive Statistics

...from previous page

- Kurtosis for a Gaussian distribution is 3.
- **Platykurtic** indicates a flattened distribution.
- **Leptokurtic** indicates a peaked distribution.
- **Mesokurtic** indicates a Gaussian, or normal, distribution.
- The terms **Gaussian** and **normal** are interchangeable.

EXERCISES

1. Using the method developed in Exercise 6.2, merge the three data files created in the Chapter 7 exercises. Change the data description to composite, and store the file as ex8__1. Load the Descript template, and File Load ex8__1. Print the summary and distribution reports on a single page.

2. Based on the distribution report, what percent of the students are 58" or less in height? Based on the mean and standard deviation, what percentage of the population is 58" or less in height?

3. Is this sample normally distributed?

Answers

1. Figure 8.7 shows the result of these operations.

Descriptive Statistics 8

```
            STATISTICAL MODELLING WITH LOTUS 1-2-3

              DESCRIPTIVE STATISTICAL ANALYSIS

         Variable: Height   Inches    Quantity:       58
             File: Ex8_1

                           N Weight              N-1 Weight
              Mean: . . . . .55.2413793 . . . . .55.2413793
           Std Dev: . . . . .2.94963670 . . . . .2.97539821
          Variance: . . . . .8.70035671 . . . . .8.85299455
           Minimum: . . . .         43  . . . . . . . . . .
           Maximum: . . . .         62  . . . . . . . . . .
             Range: . . . .         19  . . . . . . . . . .
          Skewness: . . . . .-1.1516902 . . . . . . . . . .
          Kurtosis: . . . . .3.61594556 . . . . . . . . . .

             Descr: Composite
          Operator: Joe W.              Date:   31-Jan-81
**************************************************************
   Cell                        10%  20%  30%  40%  50%  60%  70%  80%
  Limit Count    %     Cum      |    |    |    |    |    |    |    |
    40    0    0.0    0.0   |.
    41    0    0.0    0.0   |.
    42    0    0.0    0.0   |.
    43    1    1.7    1.7   |.
    44    0    0.0    1.7   |.
    45    0    0.0    1.7   |.
    46    0    0.0    1.7   |.
    47    0    0.0    1.7   |.
    48    0    0.0    1.7   |.
    49    1    1.7    3.4   |.
    50    0    0.0    3.4   |.
    51    3    5.2    8.6   |++
    52    3    5.2   13.8   |++
    53    4    6.9   20.7   |+++
    54   11   19.0   39.7   |+++++++++
    55    4    6.9   46.6   |+++
    56   10   17.2   63.8   |++++++++
    57   10   17.2   81.0   |++++++++
    58    6   10.3   91.4   |+++++
    59    3    5.2   96.6   |++
    60    1    1.7   98.3   |.
    61    0    0.0   98.3   |.
    62    1    1.7  100.0   |.
    63    0    0.0  100.0   |.
    64    0    0.0  100.0   |.
    65    0    0.0  100.0   |.
    66    0    0.0  100.0   |.
    67    0    0.0  100.0   |.
    68    0    0.0  100.0   |.
    69    0    0.0  100.0   |.
    70    0    0.0  100.0   |.
Low cell contains points below distribution
Pts above High cell        0
Total Points in dist      58
```

Figure 8.7 Answers to Exercise 8.1

8 Descriptive Statistics

2. The cumulative percent of students in cells 58 and below is 91.4%, based on the table. Solving for z:

$$z = (58 - 55.2413793)/2.97539821 = .927$$

This means that 58 is .927 standard deviations above the mean. Table A.1 in Appendix A indicates an area of probability of 0.324 for 0.93 standard deviations. Since the values below the mean are all less than 58, an additional 0.5 is added. Based on statistics, 82.4% of the students in the population are less than 58" tall.

3. The third moment about the mean — or skewness — is -1.1516902. This number indicates a slightly high bias, which is confirmed by the fact that only 46.6% of the students are less than the mean, 55". A kurtosis of 3.6 indicates a more peaked than normal distribution. There are two distinct peaks, indicating a possibly multimodal distribution, which is not surprising considering the first of the three data groups was smaller than the second and third. Clearly, this distribution is not purely normal, and caution should be exercised when applying the statistical results. The next chapter provides a more conclusive test for data normality.

TEMPLATE LISTING

The following listing was transferred from a working copy of the Descript spreadsheet into a word processor. If you are entering this program manually, use this listing instead of the segments included in the chapter. They are provided for clarification of the operation of the spreadsheet and differ slightly from this listing.

Range identifiers are used to avoid listing columns of identical cells,. A20:-H20: indicates the contents shown should be entered into cell A20 and copied over the range A20.H20.

Descriptive Statistics 8

```
A20:-H20: '**********          (Note: 11 *'s - copy over range)
A21: 'Label:
A23: 'File:
A26: 'Variable:
A31: 'Graph:
A36: 'Data:
B2:  '    DESCRIPTIVE STATISTICAL ANALYSIS
B4:  'Variable:
B5:  "File:                    (Note: Right justified)
B8:  "Mean:                         "     "     "     "
B9:  "Std Dev:                      "     "     "     "
```

Spreadsheet Labels and Marquis Listing

```
B10: "Variance:                     "     "     "     "
B11: "Minimum:                      "     "     "     "
B12: "Maximum:                      "     "     "     "
B13: "Range:                        "     "     "     "
B14: "Skewness:                     "     "     "     "
B15: "Kurtosis:                     "     "     "     "
B17: "Descr:                        "     "     "     "
B18: "Operator:                     "     "     "     "
B22: 'Dmnsion
B23: 'Name
B24: 'Descrip
B25: 'Operator
B26: 'Name
B27: 'Units
B28: 'Hi Limit
B29: 'Lo Limit
B30: 'Count
B31: 'Title-1
B32: 'Title-2
B33: 'X-title
B34: 'Y-title
B35: 'Flt Lbl
B36: 'Start
C21: 'Data:
D7:  ^N Weight
E4:  'Quantity:                (Note:/RLC - centered text)
E18: "Date:
F7:  'N-1 Weight               (Note: Right justified)
```

continued...

8 Descriptive Statistics

...from previous page

```
F18: (D1) @NOW              (Note:/RFD - date format)
I21: 'MACROs
K15:'       DESCRIPTIVE STATISTICS TEMPLATE
C8:-C15: '. . . . .        (Note: Copy over indicated ranges)
E8:-E15: '. . . . .              "         "         "         "
F11:-F15: ' . . . . .            "         "         "         "
R3:-R10: '. . . . .              "         "         "         "
T3:-T10: '. . . . .              "         "         "         "
U6:-U10: ' . . . . .             "         "         "         "
D8:-D15: '_____             "         "         "         "
F8:-F10: '_____             "         "         "         "
S3:-S10: '_____             "         "         "         "
U3:-U5:  '_____             "         "         "         "
Q19:-X19: '********
Q18: 'Print Constants
R2: 'Erase data range
R13: 'Distribution constants
R14: 'Lo Cell
R15: 'Cell size
R21: '      STATISTICAL MODELLING WITH LOTUS 1-2-3
I32: '     Descriptive Statistics Template
Y1: "Cell
Y2: "Limit
Y36: 'Low cell contains points below distribution
Y37: 'Pts above High cell
Y38: 'Total Points in dist
Z2: 'Count
AA2: "%
AB2: "Cum
AD1: '    10%   20%   30%   40%   50%   60%   70%   80%
AD2: '     |     |     |     |     |     |     |     |
```

continued...

...from previous page

```
J3:  '****************************************************
J4:  '*****************************************************
J5:  '***                                               ***
J6:  '***                                               ***
J7:  '***      STATISTICAL MODELING WITH LOTUS 1-2-3    ***
J8:  '***              COMPANION DISKETTE               ***
J9:  '***                                               ***
J10: '***                                               ***
J11: '***        A STATISTICAL SOFTWARE PACKAGE         ***
J12: '***                                               ***
J13: '***                                               ***
J14: '***                                               ***
J15: '***                                               ***
J16: '***                                               ***
J17: '***                                               ***
J18: '*****************************************************
J19: ' ****************************************************
```
Spreadsheet Labels and Marquis Listing

```
I23: 'start
I24: '/wtc{home}{goto}i1~
I25: 'Press RETURN to continue ~a1~
I26: '{home}{goto}c8~
I27: '/wtb~
I28: '/xmmenu1~
```
Autostart MACRO

```
I34: 'Menu1
I35: 'File
I36: 'Load  Unload  Save  Directory  Quit
I37: '/xmmenu2~
J35: 'Print
J36: 'Summary  Distribution  FormFeed  Quit
J37: '/xmmenu3~
K35: 'Graph
```

continued...

8 Descriptive Statistics

...from previous page

```
K36: 'Histogram  Save  Replace  Quit
K37: '/xmmenu4~
L35: 'Cursor
L36: 'Left  Right  Up  Down  GoTo  Home  Quit
L37: '/xmmenu5~
M35: 'Adjust
M36: 'Recalculate  Plot  Bin-Limits  Quit
M37: '/xmmenu6~
N35: 'Quit
N36: 'Retrieve Menu Template
N37: '/fdb:\~
N38: '/frmenu~

I40: 'Menu2 - File
I41: 'Load
I42: 'Load a file into the template
I43: '/xgload~
J41: 'Unload
J42: 'Remove a file from the template
J43: '/xgunload~
K41: 'Save
K42: 'Store the current data file
K43: '{goto}c22~/fxv{?}~.{pgdn}{down}{end}{down}~
K44: '/xmmenu2~
L41: 'Directory
L42: 'Change the directory
L43: '/fd{?}~
L44: '/xmmenu1~
M41: 'Quit
M42: 'Return to the Primary Menu
M43: '/xmmenu1~

I46: 'Menu3 - Print
I47: 'Summary
I48: 'Print the summary report
I49: '/xcsummary~
I50: '/xmmenu3~
J47: 'Distribution
```

continued...

Descriptive Statistics 8

...from previous page

```
J48:  'Print Frequency Distribution table
J49:  '/xcdist~
J50:  '/xmmenu3~
K47:  'FormFeed
K48:  'Advance page to top of form
K49:  '/pppq
K50:  '/xmmenu3~
L47:  'Quit
L48:  'Return to main menu
L49:  '/xmmenu1~

I52:  'Menu4 - Graph
I53:  'Histogram
I54:  'Create a Histogram graph
I55:  '/xchistogr~
I56:  '/xmmenu4~
J53:  'Save
J54:  'Create a PIC file from the current graph
J55:  '/gs{?}~q
J56:  '/xmmenu4~
K53:  'Replace
K54:  'Save an update to an existing PIC file
K55:  '/gs{?}~rq
K56:  '/xmmenu4~
L53:  'Quit
L54:  'Return to main menu
L55:  '/xmmenu1~
I58:  'Menu5 - Cursor
I59:  'Left
I60:  'Move cursor left five columns
I61:  '{left}{left}{left}{left}{left}
I62:  '/xmmenu5~
J59:  'Right
J60:  'Move cursor right five columns
J61:  '{right}{right}{right}{right}{right}
J62:  '/xmmenu5~
K59:  'Up
K60:  'Move cursor up one page
K61:  '{pgup}
K62:  '/xmmenu5~
```

continued...

8 Descriptive Statistics

...from previous page

```
L59:  'Down
L60:  'Move cursor down one page
L61:  '{pgdn}
L62:  '/xmmenu5~
M59:  'GoTo
M60:  'Go to a specific cell by name or number
M61:  '/wtc
M62:  '{goto}{?}~
M63:  '/xmmenu5~
N59:  'Home
N60:  'Home cursor
N61:  '{home}
N62:  '{goto}c8~/wtb
N63:  '/xmmenu5~
O59:  'Quit
O60:  'return to main menu
O61:  '{home}
O62:  '{goto}c8~/wtb
O63:  '/xmmenu1~

I65:  'Menu6 - Adjust
I66:  'Recalculate
I67:  'Recalculate spreadsheet after halt or data change
I68:  '/xg\r~
J66:  'Plot
J67:  'Replot Histogram after limits change
J68:  '/xgplot~
K66:  'Bin-limits
K68:  '/xgbinlim~
L66:  'Stop
L67:  'Halt the MACRO to make changes
L68:  '/wtc{home}
L69:  '/xq
M66:  'Quit
M67:  'Return to primary menu
M68:  '/xmmenu1~
```

Menu MACROs

continued...

Descriptive Statistics 8

...from previous page

```
I77:   'Load MACRO
I78:   '{goto}c22~
I79:   '/fcce{?}~
I80:   '{down}/c~c5~
I81:   '{down}/c.{down}~c17~
I82:   '{down}{down}/c~c4~
I83:   '{down}/c~d4~
I84:   '{goto}c36~
I85:   '/rnddata~/rncdata~.{end}{down}~
I86:   '{goto}f4~@count(data)~
I87:   '{goto}d8~@avg(data)~
I88:   '{down}@std(data)~
I89:   '{down}@Var(data)~
I90:   '{down}@min(data)~
I91:   '{down}@max(data)~
I92:   '{down}+d12-d11~
I93:   '/xnEnter the low cell value for distribution ~s14~
I94:   '/xnEnter the cell size ~s15~
I95:   '{goto}d36~(c36-$d$8)^2~
I96:   '{right}(c36-$d$8)^3~
I97:   '{right}(c36-$d$8)^4~
I98:   '{left}{Left}
I99:   '/c.{right}{right}~.{left}{end}{down}{right}~
I100:  '{up}@sum({down}.{end}{down})~
I101:  '/c~.{right}{right}~
I102:  '{up}@sqrt({down}/($f$4-1))~
I103:  '{edit}{calc}~/c~f9~
I104:  '{right}(({down}/$f$4)/($d$9)^3)~{edit}{calc}~/c~d14~
I105:  '{right}(({down}/$f$4)/($d$9)^4)-3~{edit}{calc}~/c~d15~
I106:  '{goto}d33~/re.{right}{right}{end}{down}~
I107:  '{home}{goto}f8~
I108:  '@avg(data)~
I109:  '{down}{down}((f9)^2)~{home}
I110:  '/xgplot~

I112:  'Plot - makes distribution and histogram tables
I113:  '/dfy3.y33~s14~s15~~
I114:  '/rncbin~y3.y33~
```

continued...

8 Descriptive Statistics

...from previous page

```
I115: '/dddata~bin~
I116: '/xmmenu1~
        Load and Plot MACRO's


N77: 'Unload MACRO
N78: '/rec4.d5~
N79: '/rec17.c18~
N80: '/ref4~/rek15~
N81: '/rec22.c35~
N82: '/redata~/re43.z35~
N83: '/cr3.u10~c8~
N84: '{home}
N85: '/xmmenu1~
        Unload MACRO


N88: 'Recalculate
N89: '{goto}c22~
N90: '/xgi80~
        Recalculate Start

N92: 'Binlim
N93: '/xnEnter lower bin limit~s14~
N94: '/xnEnter bin size~s15~
N95: '/xmmenu1~
        Adjust Bin sizes

Q26: 'Subroutine Summary
Q27: '/pplrq19.x22~g
Q28: 'ra1.h20~g
Q29: '/xr
        Print summary subroutine
```

continued...

...from previous page

```
Q34:  'Sub-MACRO Histogr
Q35:  '/gxy3.y34~
Q36:  'az3.z34~
Q37:  'tb
Q38:  'oss2~tf\c31~
Q39:  'ts\c32~
Q40:  'tx\c33~
Q41:  'ty\c34~q
Q42:  'ncone~vq
Q43:  '/xr
```
                Create Histogram subroutine - graph

```
T26:  'Subroutine dist
T27:  '/ppry1.ad33~g
T28:  'ry36.ad38~g
T29:  'rq19.x19~gpq
T30:  '/xr
```
                Print Frequency Distribution Table

```
AB37: (+Z35)
AB38: @SUM(Z3..Z34)

AA3:  (F1) (Z3/$AB$38)*100    (Note: /RFF1 - 1 decimal place)
AB3:  (F1) (AB2+AA3)          (Note: /RFF1 - 1 decimal place)
AC3:  '|
AD3:  (+) (AA3/2)             (Note: /RFF+ - Format +/- width 32)
```

After the four previous cells are entered, execute the following copy command

/cAA3.AD34~~   Copy AA3 to AD3 from AA4 to AD34

                Calculations for Frequency distribution

CHAPTER 9

GRAPHIC METHODS

9 Graphic Methods

The methods developed in Chapter 8 not only provide the user with the means to understand the limits and range of the data set; they also offer a great deal of information about how widely dispersed a group of values is expected to be. The **frequency distribution** table and **histogram** provide a graphic look at the distribution of the data. **Kurtosis** and **skewness** measure how peaked and how skewed the data are, respectively. These techniques are the backbone of statistics. The methods presented in this chapter are an enhancement, providing graphs that can be used in conjunction with the mathematical analyses.

The simplest technique in this section is simply a plot of the data values. These values can be graphed in the order they appear or sorted in ascending or descending order. To augment the frequency distribution and histogram, this chapter will introduce **ogive**—a cumulative plot of the cells in a histogram (often normalized to percent). Its most useful feature is the ability to cross-reference frequency to magnitude of the independent variable, which makes it a valuable tool for setting limits or acceptance criteria based on measured data. Using the equation for the Gaussian probability density function, a pseudo-distribution is created. This data set has the same number of points as the data under analysis. The pseudo-data are effectively the data values one would expect to see for a given mean and standard deviation. Two means are provided for graphing the actual data set and the expected data set based on the pseudo-distribution. Advanced readers can take one of these pseudo data sets and enter it into the descriptive statistics template. The skewness and kurtosis are zero.

The final section of this template performs a chi-squared test on the following hypotheses:

H_0: The sample data is from a normally distributed population.
H_a: The population is not normally distributed.

The calculations table created in Chapter 8 is expanded to include the expected values from the pseudo-distribution. From this table, a difference table is generated. The summation of the elements in this table, when weighted by the individual expected values, has a chi-squared distribution. The summation value, compared to the value from a chi-square table, is the accept/reject criteria for the test.

Graphic Methods 9

Obviously, there is much similarity between this template and the one in the previous chapter. The Descriptive Statistics template is stored under the name DESCRIPT. A copy of that template was stored under the name CHI_SQ. This copy was expanded into the template for this chapter. Both of these spreadsheets are included on the companion disk, and both can be called from the MENU template in Appendix D.

The CHI_SQ template provides all the information available in DESCRIPT. There are two reasons for preserving both entities. The first reason is simplicity. If you don't need the chi-square information, don't complicate the output with it. Also, it is simpler to train an operator to use the DESCRIPT template. The second reason for including both copies is variety. It is helpful to have two similar but slightly different outputs. Each template has uses with subtle changes in output to provide information efficiently and effectively.

Figure 9.1 is a copy of Figure 8.1b with a **type-two Gaussian** plot substituted for the histogram in Chapter 8. In this graph, the histogram's information is still present, but the graph also includes the curve of the normally distributed pseudo-distribution. Clearly, the expected value curve has the characteristic "bell" shape. The actual data points have a similar shape, but every other value is zero, which is due to an error in selecting the cell size. If the cell is smaller than the measurement size, false information can result. This is often called a granularity error. A more extreme example would be measurements to the nearest .5" with cell size of .3". Every third cell would be zero.

There is a significant difference between these two curves, resulting in an artificially high chi-square value. A corrected curve will be shown later in this section. Examination of the frequency distribution table after data are plotted allows adjustments to granularity to be made quickly and easily. Always take care when selecting cell limits.

9 Graphic Methods

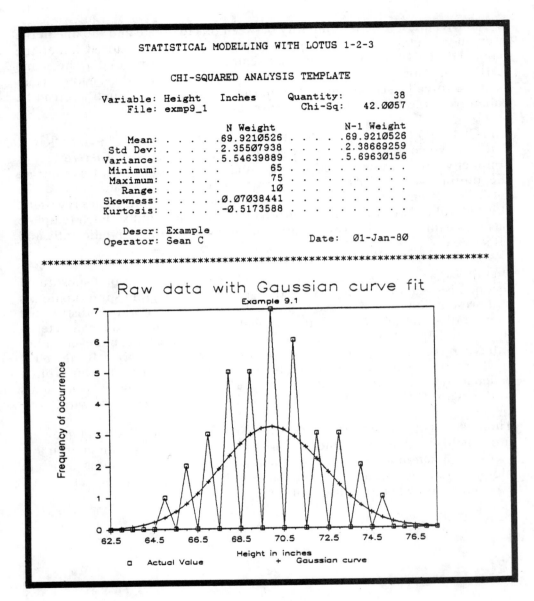

Figure 9.1 Sample Output

Graphic Methods 9

The contents of this template are as follows:

- Data inputs are similar in nature to those for Chapter 8, but should be normally distributed.

- All information from the DESCRIPT template is included.

- Outputs include the following:

  1. Data graph

  2. Ogive (graph)

  3. Actual vs. expected value bar graph

  4. Actual vs. expected value line graph

  5. Chi-squared coefficient of the difference between actual and expected values

MENUS

The menu structure for this template is shown in Figure 9.2. The sections in boldface type are the additions to the original spreadsheet that constitute the CHI_SQ template. Only these sections are described in detail in this chapter.

9 Graphic Methods

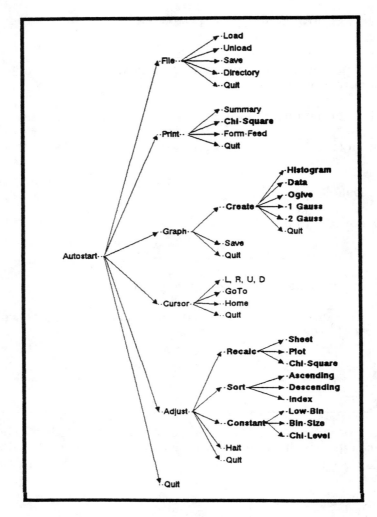

Figure 9.2 Chi-squared Template Menu

Print

The only change in the Print menu is the substitution of chi-square for the distribution table created in Chapter 8. There are only two differences between Figures 9.1 and 8.1. In Figure 9.1, the chi-square table has the coefficient in the upper right corner and a column after cumulative percent that gives the expected values.

Graph

The **Graph** menu is expanded to two levels. The histogram graph from Chapter 8 is now located in the **Create** submenu with the other graphs to be created.

The **data** graph is a simple line plot of the values in the spreadsheet data range. The elements are indexed and will be plotted in the order in which they appear when the graph is generated. The labels for this graph are those stored with the data file.

The **ogive** is a two-way graph. The X-axis is the column of cell limits in the frequency distribution table. The A range is the cumulative percent column.

Gauss 1 and **Gauss 2** are identical except that 1 is a bar graph and 2 is a line graph. Two ordinate scales are provided. The first is the data count column from the frequency distribution table. The second is the expected value column. These curves allow the user to visually compare the test data set to a "normal" data set with the same statistical parameters.

Cursor

The Goto range on the Cursor menu is given a submenu. Home is moved to this submenu along with the chi-table and address options, which allows direct access to the table or any address and return to home. These commands are based on 1-2-3's ability to "goto" a range name directory.

Adjust

Recalculate is restructured to allow the entire spreadsheet, or just the tables, to be recalculated. If data are entered or edited, the spreadsheet recalculate command must be used. If only the bin limits or chi-square level are adjusted, the whole process need not be repeated. The **chi-table** command executes the tail end of the recalculate command, which builds the tables.

Sort is added to allow sorting of the data in column C. The data can be arranged in ascending or descending order or returned to their original order.

Constants that can be changed are the low bin in the distribution, bin size, or the significance level for the chi-square test.

Halt stops macro execution to allow changes.

9 Graphic Methods

METHODS AND APPLICATIONS

Comparing Figure 8.3 with 9.3 underscores the rationale behind adding the current material to the previous template. The five new menus described in Figure 9.2 are added in the third column on the block diagram. All information needed for generating the graphs and results in this section was previously generated in the descriptive statistics template. Both templates are included to allow variation in outputs and to simplify the introduction of the material. Most people will only use the version created in this chapter.

Note: equations 9.1-9.2a are included in Figure 9.4; equations 9.3 and 9.3a are included in Figure 9.5.

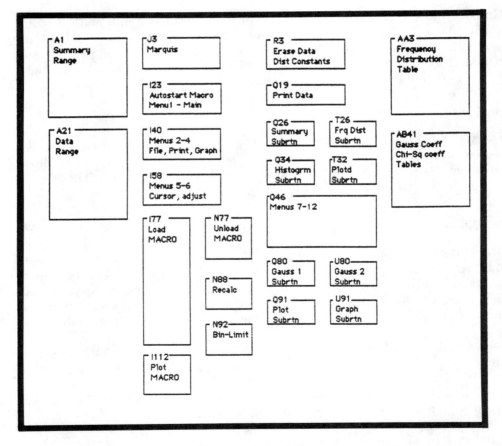

Figure 9.3 Chi-squared Analysis Template Block Diagram

Graphic Methods **9**

The spreadsheet block ranging from AB43 to AB73 is a column of coefficients based on Equation 9.1. This will be recognized as the equation for the magnitude of the Gaussian **probability density function**, or P.D.F. (This is the equation for the classic "bell-shaped" curve.) The particular form chosen here provides the ordinates based on a specified mean and standard deviation, which you have already calculated. Equation 9.1a restates the equation in the form used in the spreadsheet.

9.1 $$Y = \frac{\exp^{-(x-\mu)^2/2\sigma^2}}{\sigma\sqrt{2\pi}}$$

9.1A $$\frac{(\text{@exp}((-(y3-\$d\$8)\wedge 2)/(2*(\$d\$9\wedge 2))))}{\$d\$9*\text{@sqrt}(2*\text{@pi})}$$

9.2 $\quad Ce = Cl * P / S$

9.2A $\quad (AB43 / \$AB\$74) * \$F\$4$

Ce = Estimated # points in cell

Cl = Magnitude of coefficient cell

P = Number of points in sample

S = Sum of the ordinates

Figure 9.4 Gaussian Equations

9 Graphic Methods

The P.D.F. equation (Equation 9.1) is entered into the first cell of the range. Some variables require relative cell addresses, while others must be absolute addresses. The formula is copied to the 30 cells below it. The result is a column of magnitudes whose amplitudes are based on Equation 9.1. Mean and deviation are from absolute cell locations, and x values are from the cell limits column of the histogram. At the bottom of the column is a summation of the ordinate values used to scale this curve to the present data set. Dividing any single element in this range by the total yields the probability that a sample will fall in that range. If the probability found this way is .35 (35%), then, for a sample population of 100 elements, the expected value is

$$Ev = 100 * 0.35 = 35$$

The table of expected values corresponding to the existing bin limits is built in this manner.

Equation 9.2 generates a value for a given cell limit that is equal to the number of points expected to be in the cell (if the data are normally distributed). When this equation is copied over the range AC3:-AC33:, it adds the expected values column to the frequency distribution column. When the values in this column are fed into the descriptive statistics template as data, the mean and standard deviation of the resulting pseudo-data set are equal to those used to generate the expected values. The kurtosis and skewness, though, are both zero.

Equation 9.3 provides the chi-square coefficient for any cell in the table. Its value is the square of the difference between actual and expected values, divided by the expected value of the cell. These coefficients have a chi-squared distribution, so their sum can be compared to the critical value from a chi-squared table, for a given confidence level and degree of freedom, to satisfy the following hypothesis:

H_0: The sample comes from a normally distributed population.

The chi-square coefficient is copied to a cell on the output to simplify hypothesis testing. After data are entered, the value is compared to the chi-square table. If the calculated coefficient is less than the critical value, then the population can be considered Gaussian (normal).

$$9.3 \quad X^2 = \sum \frac{(Fo - Fc)^2}{Fc}$$

9.3A   ((Z3 - AC3)^2)/AC3

X^2 = Chi-Squared value

Fo = Observed # of values

Fc = Expected number of values

Figure 9.5 Chi-squared Equations

OUTPUTS

The **adjust constants** option, followed by the **adjust plot** option, allows the user to change the minimum cell or cell size. Figure 9.6 is a replot of the data used to generate Figure 9.1, using a cell size of 1" instead of 0.5". Two things have happened: the cells now are packed with no spaces, and the chi-squared value is reduced from 42.00 to 4.166. The example figures for the next section were generated using this adjusted data set.

9 Graphic Methods

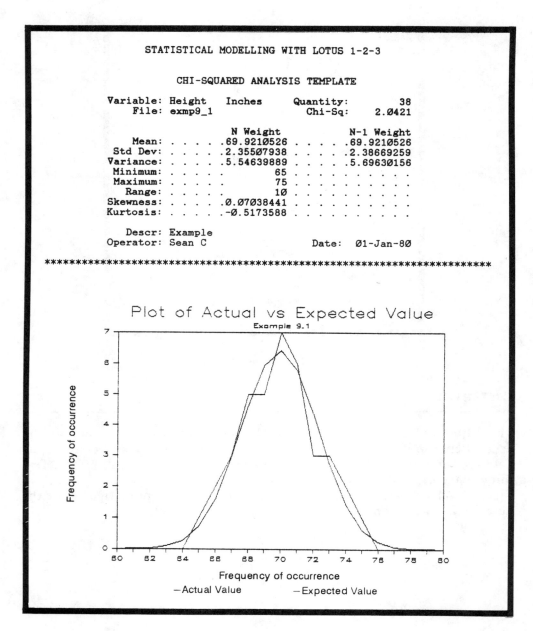

Figure 9.6 Replot of Figure 9.1

Graphic Methods 9

Four graph subroutines have been added to accomplish the outputs for this template: **Plotd**, **Ogive**, **Gauss 1**, and **Gauss 2**. In addition to these four graphs is the ammended frequency distribution table, now called the chi-square table, which provides the coefficient for the hypothesis test.

Plotd generates a line graph of the data values in the range named data, in their present order (see Figure 9.7). The labels on the graph are those stored and loaded with the data file. The X-axis on this graph is an index column added to the right of the data; the Y-axis is the magnitude of the data. This plot allows a quick look at a data set in the order it was entered or sorted into ascending or descending order, which can serve as a source for ideas about what analysis to use or what trends exist in the data.

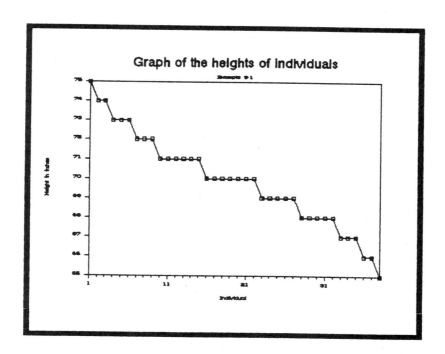

Figure 9.7 Line Graph of Data Values

9 Graphic Methods

The second graph routine creates an ogive. In this line graph, the cell limit is plotted concurrently with the cumulative cell count. If all data points are within the range limits of the frequency distribution table, the ogive runs from 0 to 100%. The characteristic curve of the ogive lends itself to quick cross-referencing of data sets. What percent of a tested population falls below a given limit can be easily identified, thus providing an empirical analysis based on the data set. In the sample data set in Figure 9.8, the top 10% of the population's height can be projected by moving horizontally across the 90% line to the curve and then down to 73.5". Three of the 38 points in this sample fall into this category.

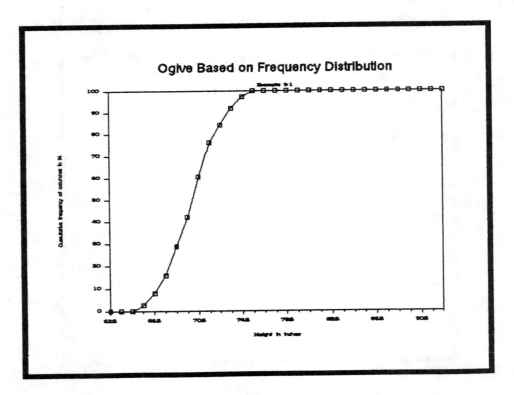

Figure 9.8 Ogive — Cumulative Cell Count

Graphic Methods 9

The top title line and Y-axis title on the ogive are fixed. The X-axis title is the Y-axis title from the line plot, which works because the ogive is generated from a frequency distribution table that has the same number of units as the population from which it is extracted. The second title line at the top is the same as the second title line on the line plot.

Two Gaussian plots are generated: one in bar form, and the other in line form. Each of these plots has the same data shown in the frequency distribution, displayed concurrently with the expected values from the pseudo-distribution. This concurrent display is valuable because a statistically correct sample taken from a normal distribution resembles the graph of the expected values. Figure 9.1 has already demonstrated how the quality of the data set can be quickly checked with this format. Granularity errors stand out boldly.

Figures 9.9 and 9.10 are Gaussian plots of the example data. As with the ogive, the top line and Y-axis are fixed. The second title line is the same as the plotd and ogive, direct from the data set. The X-axis title is the Y-axis title from the data set.

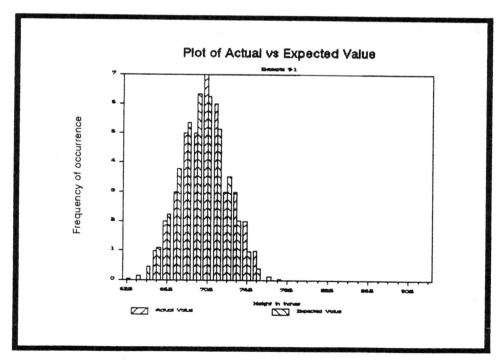

Figure 9.9 Bar Graph of Acutal vs. Expected Value

9 Graphic Methods

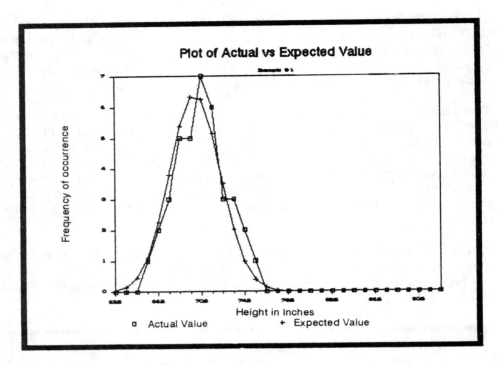

Figure 9.10 Line Graph of Acutal vs. Expected Value

As in Chapter 8, the outputs can be mixed. The summary table can be printed alone or with the chi-square table. Any one (or more) graph(s) can be printed with the summary.

COMMANDS USED

Following is the list of commands used in the template created in this chapter:

{edit}{calc}	Converts cell contents to value
@std()	Computes standard deviation
@today	Prints today's date (CPU clock)
@var	Computes variance
/fcce{?}~	Merges file into spreadsheet
/fdb:\~	Changes default directory to B:

continued...

Graphic Methods **9**

...from previous page

/fx{?}~	eXtracts a file or segment
/gs{?}~rq	Saves a graph
/rncdata~	Creates a range named data
/rnddata~	Deletes a range named data
/xc	Macro gosub
/xg	Macro Goto
/xl	Macro label input
/xn	Macro number input
/xr	Macro returns from subroutine
/xm	Macro menu
/wtb	Sets both titles
/wtc	Clears titles

MACRO FUNCTIONS

The first line in the subroutine for generating the line plot moves the cursor to the top value in the label range, which is necessary to allow the relative addressing required to set the X range. The second line executes a /grg (Graph Reset Graph), which clears all graph settings. If this command is not executed, when the titles are set using the \address option, the title line becomes \address\adress, and it no longer functions correctly. Continuing in cell T34, the X range is set to the label range using ".{end}{down}"; a range is set to "data" (the named data range); type is set to "line" (tl), and the top title line is set to the contents of cell C31 (otf\C32 options title first \C31).

Cell T35 sets the second title line (ts\C32), the X-axis (tx\C33), and the Y-axis (ty\C34). Scale skip is set to 10 (ss10) to prevent overlapping of numbers on the X scale, and options mode is quit and viewed (qv). Line T36 returns from the subroutine.

Line Graph of Data Values

```
T32:   'Subroutine plotd
T33:   '{goto}d36~
T34:   '/grgx.{end}{down}~adata~tlotf\c31~
T35:   'ts\c32~tx\c33~ty\c34~ss10~qvq
T36:   '/xr
```

9 Graphic Methods

The other four graph subroutines work in an identical fashion and will not be explained separately.

There are three parts to the Graph Save subroutine. Cells U91 to U94 clear the worksheet titles and home the cursor. At cell C8, a horizontal window is installed. In U93, the window command ({window}) moves the cursor into the lower section of the display. The Goto command ({goto}I106) moves the display to the save message that provides operator instructions. U94 sets the spreadsheet into graph mode.

Graph Save Subroutine (Part I)

```
U91:    'Save                               (Saves graph)
U92:    '/wtc{home}{goto}c8~
U93:    '/wwh{window}{goto}I106~
U94:    '/g
```

After the previous section has been executed, the macro halts when it encounters a cell with no command under it, thus allowing the user to alter the graph to suit the particular needs of the project at hand. Scales can be set manually with grids installed, labels changed, etc. Any adjustments can be made. The user is prompted to save the graph when adjustments are complete.

When the restart command (\g) is issued, the macro at U96 is executed. The lower window is cleared (/wwc) titles are reinserted, and control is returned to menu 4, the Graph menu.

Graph Save Subroutine (Part II)

```
U96:    'Restart - using \g
U97:    '/wwc{home}{goto}c8~/wtb
U98:    '/xmmenu4~
```

The following cells (K106-110) are the Operator Instructions menu for the Graph Save subroutine:

Graph Save Subroutine (Part III: Operator Instructions Menu)

```
K106:  'Enter required adjustments to graph
K107:  'Enter (S)ave
K108:  'Enter Graph name
K109:  'Quit Graph mode (q)
K110:  'Press altG to Restart
```

The commands in the following cells do not actually constitute a macro, but they generate the chi-square table. Column AA is the data from column z, bin counts, normalized to percent of the total number of points. AB is the cumulative percent. AC is the expected value found by dividing the corresponding cell in the range beginning at AB43 by the sum of the range to determine the probability that a certain value exists and then multiplying by the total number of values. AD is a label range to improve readability of the table. Values in AE are half those in AA, but the cell is of the +/- format. When the cell width is extended to 42 columns, the histogram prints to the right of the table.

The ranges starting at AB and AE43 are the expected value and chi-square coefficients.

Calculations for Frequency Distribution

```
AA3:   (F1) (Z3/$AC$38)*100                  (Copy AA3 to range AA3.AA34)
AB3:   (F1) (AB2+AA3)                        (Copy AB3 to range AB3.AB33)
AC3:   (F1) (AB43/$AB$74)*$F$4'              (Copy from AC3 to range AC3.AC33)
AD3:   '|                                    (Copy from AD3 to range AD3.AD34)
AE3:   (+) (AA3/2)                           (Copy from AE3 to range AE3.AE34)
AB43:  (F2)(@EXP((-DDY3$D$8)^2)/(2*($D$9^2))))/$D$9*@SQRT(2*@PI)
                                             (Copy AB43 to range AB43.AB73)
AE43:  ((Z3-AC3)^2)/AC3                      (Copy from AE43 to range Ae43.AE73)
```

The Binlim subroutine allows update of the constants used to plot the frequency distribution.

9 Graphic Methods

Binlim Subroutine

N87: 'Binlim
N88: '/xnEnter lower bin limit~s14~
N89: '/xnEnter bin size~s15~
N90: '/xmmenu1~

RANGE NAMES

After the spreadsheet information is entered, some final adjustments are needed to make it perform. The following list should be entered as shown. The first group of commands contains range names for the macro Goto commands; the second group contains column width adjustments.

```
/rncbinlim<return>n93<return>
/rnchistogr<return>q35<return>
/rncload<return>i78<return>
/rncmenu1<return>i35.n36<return>
/rncmenu2<return>i41.m42<return>
/rncmenu3<return>i47.l48<return>
/rncmenu4<return>i53.l54<return>
/rncmenu5<return>i59.o60<return>
/rncmenu6<return>i66.m67<return>
/rncplot<return>i113<return>
/rncsummary<return>q27<return>
/rnctable<return>y1<return>
/rncunload<return>n78<return>
/rnc\0<return>i24<return>
/rnc\a<return>i26<return>
/rnc\r<return>n89<return>
/rf+ae3.ae34<return>
```

For each of the following columns, perform the following steps:

- Move the cursor to the first row in the column.
- Execute /wcs(enter the number) <return>.

Column	Width
A	10
D	11
F	11
H	3
Z	5
AA	6
AB	7
AC	1
AD	44

SUMMARY

- Graphic methods augment mathematical methods to provide a more complete understanding of data.

- A **chi-square** hypothesis is a test to determine whether a data set is from a normal distribution.

- Comparison of a **histogram** to the expected values of a **Gaussian** data set with identical mean and standard deviation is an excellent visual test for normality.

- **Kurtosis** and **skewness** are both zero for a set of expected values since, by definition, they are Gaussian.

- An **ogive** relates bin sizes to cumulative percent in a characteristic curve useful for making projections based on empirical data.

EXERCISES

1. Using the data file created in Exercise 8.1 (from Chapter 8), print a summary table and chi-square table using template CHI__SQ. Based on the chi-square statistic, is this distribution normal?

9 Graphic Methods

2. Generate a histogram graph. When the graph is saved, change the range so that the histogram fills the display range.

3. Change the low bin to make the top 19% of data values fall out of range to the top. Generate an ogive. What has happened to the Y scale?

4. Restore the bin to its original value. Replot the spreadsheet. Using the ogive, determine what elements need to be removed to eliminate the upper and lower 10% of values. Edit these out of the data range. (Hint: use adjust halt and manually edit the file.) Recalculate the spreadsheet. What has happened to the chi-square coefficient?

5. Repeat Exercise #4, removing 36% from the top and bottom of the original distribution. What happened to the chi-square coefficient? Why did this happen?

Answers

1. Figure 9.11 shows the result of this operation. There are 58 data points in this set, so there are 56 degrees of freedom. Using the chi-square table in Appendix A, and a 0.05 level, the critical value for chi-squared is 79.082. In this example, the value was rounded to 60. The computed value for chi-squared is 711, which is significantly larger than 79. The distribution is not normal.

Graphic Methods 9

```
            STATISTICAL MODELLING WITH LOTUS 1-2-3

                    CHI-SQUARED ANALYSIS TEMPLATE
         Variable: Height    Inches     Quantity:        58
             File: EX8_1                  Chi-Sq:   711.0888

                            N Weight                N-1 Weight
              Mean: . . . . .55.2413793  . . . . . .55.2413793
           Std Dev: . . . . .2.94963670  . . . . . .2.97539821
          Variance: . . . . .8.70035671  . . . . . .8.85299455
           Minimum: . . . . .         43 . . . . . . . . . . .
           Maximum: . . . . .         62 . . . . . . . . . . .
             Range: . . . . .         19 . . . . . . . . . . .
          Skewness: . . . . .-1.1516902  . . . . . . . . . . .
          Kurtosis: . . . . .3.61594556  . . . . . . . . . . .

             Descr: Composite
          Operator: Jow W.              Date:    01-Jan-80
***************************************************************
 Cell
Limit Count   %    Cum    EV    10%  20%  30%  40%  50%  60%  70%  80%
  40    0   0.0    0.0    .0   |.     |    |    |    |    |    |    |
  41    0   0.0    0.0    .0   |.
  42    0   0.0    0.0    .0   |.
  43    1   1.7    1.7    .0   |.
  44    0   0.0    1.7    .0   |.
  45    0   0.0    1.7    .0   |.
  46    0   0.0    1.7    0.1  |.
  47    0   0.0    1.7    0.2  |.
  48    0   0.0    1.7    0.4  |.
  49    1   1.7    3.4    0.8  |.
  50    0   0.0    3.4    1.6  |.
  51    3   5.2    8.6    2.3  |++
  52    3   5.2   13.8    4.3  |++
  53    4   6.9   20.7    5.9  |+++
  54   11  19.0   39.7    7.2  |+++++++++
  55    4   6.9   46.6    7.8  |+++
  56   10  17.2   63.8    7.6  |++++++++
  57   10  17.2   81.0    6.6  |++++++++
  58    6  10.3   91.4    5.1  |+++++
  59    3   5.2   96.6    3.5  |++
  60    1   1.7   98.3    2.1  |.
  61    0   0.0   98.3    1.2  |.
  62    1   1.7  100.0    0.6  |.
  63    0   0.0  100.0    0.2  |.
  64    0   0.0  100.0    0.1  |.
  65    0   0.0  100.0    .0   |.
  66    0   0.0  100.0    .0   |.
  67    0   0.0  100.0    .0   |.
  68    0   0.0  100.0    .0   |.
  69    0   0.0  100.0    .0   |.
  70    0   0.0  100.0    .0   |.
Low cell contains points below distribution
Pts above High cell         0
Total Points in distrib.   58
```

Figure 9.11 Answer to Exercise 9.1

9 Graphic Methods

2. The histogram is generated by entering the Graph Create menu. After these commands are entered, select the X range required to fill the screen with the plot. In this case, it is 42-64 inches. Enter (O)ptions (S)cale (Y)-axis (M)anual. Set (L)ower to 42 and (U)pper to 64. Press QQV and save the graph as exer9_2. Figure 9.12 shows this graph. Quit Graph mode, and press <Alt/G> to restart the macro. Quit the Graph macro menu.

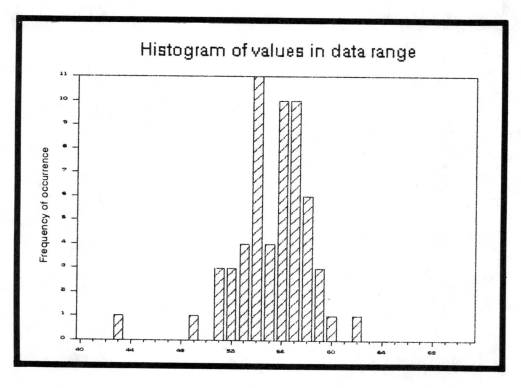

Figure 9.12 Answer to Exercise 9.2

3. From the main menu of this template, reload this spreadsheet. Based on the cumulative column in Figure 9.11, cell 57 should be the top of the new distribution. Enter a low cell of 27 and a cell size of 1. Execute a Graph Create ogive. The ogive created does not go all the way to the top of the chart. Note that the chi-square value has been significantly reduced.

4. Removing the outlying values that distorted the distribution has made it closer to being normally distributed.

5. The chi-square value has again jumped up since the truncated distribution is less normal than before.

TEMPLATE LISTING

The following template listing provided here was transferred from a working spreadsheet by printing it to a file from Lotus 1-2-3, using options unformatted and cell contents. The file was then imported into a spreadsheet, sorted, and printed to a file for transfer to a word processor. It contains all of the functions from the template in Chapter 8 plus the additional functions generated for Chapter 9.

```
A20:-H20: '**********

A21:  'Label:
A23:  'File:
A26:  'Variable:
A31:  'Graph:
A36:  'Data:

B2:   '              CHI-SQUARED ANALYSIS TEMPLATE
B4:   'Variable:
B5:   "File:
B8:   "Mean:
B9:   "Std Dev:
B10:  "Variance:
B11:  "Minimum:
B12:  "Maximum:
B13:  "Range:
B14:  "Skewness:
B15:  "Kurtosis:
B17:  "Descr:
B18:  "Operator:
B22:  'Dmnsion
B23:  'Name
B24:  'Descrip
```

continued...

9 Graphic Methods

...from previous page

```
B25: 'Operator
B26: 'Name
B27: 'Units
B28: 'Hi Limit
B29: 'Lo Limit
B30: 'Count
B31: 'Title-1
B32: 'Title-2
B33: 'X-title
B34: 'Y-title
B35: 'Flt Lbl
B36: 'Start

C8:-C15: '. . . . .
C21: 'Data:

D7: ^N Weight
D8:-D15: '_____

E4: 'Quantity:
E5: "  Chi-Sq:
E8:-E15: '. . . . .
E18: "Date:

F5: @SUM(AE43..AE73)
F7: 'N-1 Weight
F8:-F10: '_____
F11:-F15: ' . . . . .
F18: (D1) [W11] @NOW

R2: 'Erase data range
R3:-R10: '. . . . .
T3:-T10: '. . . . .
U6:-U10: ' . . . . .
S3:-S10: '_____
U3:-U5: '_____
Q19:-X19: '********
R21: '      STATISTICAL MODELLING WITH LOTUS 1-2-3
```

continued...

Graphic Methods 9

...from previous page

```
Q18: 'Print Constants
R13: 'Distribution constants
R14: 'Lo Cell
R15: 'Cell size
R16: 'Chi-sq
S14: (D11-(5*D13/20))
S15: (D13/20)
S16: 0.9


Y1:  "Cell
AE1: '     10%   20%   30%   40%   50%   60%   70%   80%
Y2:  "Limit
Z2:  'Count
AA2: "%
AB2: "Cum
AC2: "EV
AE2: '      |     |     |     |     |     |     |     |


J3:  '***********************************************************
J4:  '***********************************************************
J5:  '***                                                     ***
J6:  '***                                                     ***
J7:  '***          STATISTICAL MODELING WITH LOTUS 1-2-3      ***
J8:  '***                  COMPANION DISKETTE                 ***
J9:  '***                                                     ***
J10: '***                                                     ***
J11: '***            A STATISTICAL SOFTWARE PACKAGE           ***
J12: '***                                                     ***
J13: '***                                                     ***
J14: '***                                                     ***
J15: '***                                                     ***
K15: '          Chi-Squared Analysis Template
J16: '***                                                     ***
J17: '***                                                     ***
J18: '***********************************************************
J19: ' **********************************************************
```

Marquis and Label Ranges

continued...

9 Graphic Methods

...from previous page

```
I21: 'MACROs
I23: 'start
I24: '/wtc{home}{goto}i1~
I25: '/xlPress RETURN to continue ~a1~
I26: '{home}{goto}c8~
I27: '/wtb~
I28: '/xmmenu1~


I34: 'Menu1
I35: 'File
I36: 'Load  Unload  Save  Directory  Quit
I37: '/xmmenu2~
J35: 'Print
J36: 'Summary  Distribution  ChiSquare  FormFeed  Quit
J37: '/xmmenu3~
K35: 'Graph
K36: 'Create  File  Quit
K37: '/xmmenu4~
L35: 'Cursor
L36: 'Left  Right  Up  Down  GoTo  Quit
L37: '/xmmenu5~
M35: 'Adjust
M36: 'Recalculate  Sort  Constant  Halt  Quit
M37: '/xmmenu6~
N35: 'Quit
N36: 'Retrieve Menu Template
N37: '/fdb:\~
N38: '/frmenu~


I40: 'Menu2 - File
I41: 'Load
I42: 'Load a file into the template
I43: '/xgload~
J41: 'Unload
J42: 'Remove a file from the template
J43: '/xgunload~
K41: 'Save
```

continued...

Graphic Methods 9

...from previous page

```
K42:  'Store the current data file
K43:  '{goto}c22~
K44:  '/fxv{?}~.{pgdn}{down}{end}{down}~{?}~
K45:  '{home}/xmmenu1~
L41:  'Directory
L42:  'Change the directory
L43:  '/fd{?}~
L44:  '/xmmenu1~
M41:  'Quit
M42:  'Return to the Primary Menu
M43:  '/xmmenu1~

I46:  'Menu3 - Print
I47:  'Summary
I48:  'Print the summary report
I49:  '/xcsummary~
I50:  '/xmmenu3~
J47:  'Chi-Square
J48:  'Print the Chi-Square table
J49:  '/xcdist~
J50:  '/xmmenu3~
K47:  'FormFeed
K48:  'Advance page to top of form
K49:  '/pppq
K50:  '/xmmenu3~
L47:  'Quit
L48:  'Return to main menu
L49:  '/xmmenu1~

I52:  'Menu4 - Graph
I53:  'Create
I54:  'Histogram  Data  Ogive  1-Gauss  2-Gauss  Quit
I55:  '/xmmenu7~
J53:  'Save
J54:  'Save the current graph to disk
J55:  '/xgsave~
K53:  'Quit
K54:  'Return to main menu
K55:  '/xmmenu1~
```

continued...

255

9 Graphic Methods

...from previous page

```
I58:  'Menu5 - Cursor
I59:  'Left
I60:  'Move cursor left five columns
I61:  '{left}{left}{left}{left}{left}
I62:  '/xmmenu5~
J59:  'Right
J60:  'Move cursor right five columns
J61:  '{right}{right}{right}{right}{right}
J62:  '/xmmenu5~
K59:  'Up
K60:  'Move cursor up one page
K61:  '{pgup}
K62:  '/xmmenu5~
L59:  'Down
L60:  'Move cursor down one page
L61:  '{pgdn}
L62:  '/xmmenu5~
M59:  'GoTo
M60:  'Home   Distribution   Address
M61:  '/wtc
M62:  '/xmmenu9~
N59:  'Quit
N60:  'return to main menu
N61:  '{home}
N62:  '{goto}c8~/wtb
N63:  '/xmmenu1~

I65:  'Menu6 - Adjust
I66:  'Recalculate
I67:  'Spreadsheet   Tables
I68:  '/xmmenu10~
J66:  'Sort
J67:  'Ascending   Descending   Index
J68:  '/xmmenu11~
K66:  'Constant
K67:  'Low-Bin  Bin-Size  Chi-Level
K68:  '{goto}q12~
K69:  '/xmmenu12~
L66:  'Halt
```

continued...

...from previous page

```
L67:  'Halt the MACRO to make changes
L68:  '/wtc{home}
L69:  '/xq
M66:  'Quit
M67:  'Return to primary menu
M68:  '/xmmenu1~


Q46:  'Menu7 - Create Graphs
Q47:  'Histogram
Q48:  'Create a Histogram Graph
Q49:  '/xchistogr~
Q50:  '/xmmenu7~
R47:  'Data
R48:  'Plot Data Points Sequentially
R49:  '/xcplotd~
R50:  '/xmmenu7~
S47:  'Ogive
S48:  'Generate Ogive From Frequency Distribution
S49:  '/xcogive~
S50:  '/xmmenu7~
T47:  '1-Gauss
T48:  'Generate Gaussian Bar Graph
T49:  '/xcgauss1~
T50:  '/xmmenu7~
U47:  '2-Gauss
U48:  'Generate Gaussian Line Graph
U49:  '/xcgauss2~
U50:  '/xmmenu7~
V47:  'Quit
V48:  'Return to Primary Menu
V49:  '/xmmenu4~


Q53:  'Menu9 - GoTo Options for Cursor
Q54:  'Home
Q55:  'Home cursor and set titles
Q56:  '{home}{goto}c8~
Q57:  "/wtb
```

continued...

9 Graphic Methods

...from previous page

```
Q58:  '/xmmenu1~
R54:  'Distribution
R55:  'View Frequency Distribution Table
R56:  '{goto}table~
R57:  '/xmmenu9~
S54:  'Address
S55:  'GoTo a specific cell
S56:  '{goto}{?}~
S57:  '/xmmenu9~


Q60:  'Menu10 - Recalculate options
Q61:  'Sheet
Q62:  'Recalculate the Statistics template
Q63:  '/xg\r~
R61:  'Plot
R62:  'Replot the frequency distribution
R63:  '/xgplot~
S61:  'Chi-Square
S62:  'Recalculate the Chi-Square analysis


Q67:  'Menu11 - Sort options
Q68:  'Ascending
Q69:  'Resort data in ascending order
Q70:  '{goto}c36~/dsrd.{end}{down}{right}~
Q71:  'p.{end}{down}~a~g
Q72:  '/xmmenu1~
R68:  'Descending
R69:  'Resort data in descending order
R70:  '{goto}c36~/dsrd.{end}{down}{right}~
R71:  'p.{end}{down}~d~g
R72:  '/xmmenu1~
S68:  'Index
S69:  'Resort data to index - original order
S70:  '{goto}c36~/dsrd.{end}{down}{right}~
S71:  'p{right}.{end}{down}~a~g
S72:  '/xmmenu1~
```

continued...

Graphic Methods 9

...from previous page

```
Q74: 'Menu12 - Change Constants
Q75: 'Low-bin
Q76: 'Change the low bin for frequency distribution
Q77: '/xnEnter New Low-Bin Value  ~s14~
Q78: '/xmmenu12~
R75: 'Bin-Size
R76: 'Change the bin size for frequency distribution
R77: '/xnEnter new Bin Size  ~s15~
R78: '/xmmenu12~
S75: 'Chi-Level
S76: 'Change the Chi-square test level
S77: '/xnEnter new Chi-Square confidence level  ~s16~
S78: '/xmmenu12~
T75: 'Quit
T76: 'Return to main menu
T77: '{home}
T78: '/xmmenu6~
```

Menu Macros

```
I72: 'Load MACRO
I73: '{goto}c22~
I74: '/fcce{?}~
I75: '{down}/c~c5~
I76: '{down}/c.{down}~c17~
I77: '{down}{down}/c~c4~
I78: '{down}/c~d4~
I79: '{goto}c36~
I80: '/rnddata~/rncdata~.{end}{down}~
I81: '{goto}f4~@count(data)~
I82: '{goto}d8~@avg(data)~
I83: '{down}@std(data)~
I84: '{down}@Var(data)~
I85: '{down}@min(data)~
I86: '{down}@max(data)~
I87: '{down}+d12-d11~
```

continued...

9 Graphic Methods

...from previous page

```
I88:  '/xnEnter top of low cell ~s14~
I89:  '/xnEnter Cell Size ~s15~
I90:  '{goto}d36~(c36-$d$8)^2~
I91:  '{right}(c36-$d$8)^3~
I92:  '{right}(c36-$d$8)^4~
I93:  '{left}{Left}
I94:  '/c.{right}{right}~.{left}{end}{down}{right}~
I95:  '{up}@sum({down}.{end}{down})~
I96:  '/c~.{right}{right}~
I97:  '{up}@sqrt({down}/($f$4-1))~
I98:  '{edit}{calc}~/c~f9~
I99:  '{right}(({down}/$f$4)/($d$9)^3)~{edit}{calc}~/c~d14~
I100: '{right}(({down}/$f$4)/($d$9)^4)-3~{edit}{calc}~/c~d15~
I101: '/re.{end}{down}{left}{left}~
I102: '{home}{goto}f8~
I103: '@avg(data)~
I104: '{down}{down}((f9)^2)~{home}
I105: '/xgplot~


N72:  'Unload MACRO
N73:  '/rec4.d5~
N74:  '/rec17.c18~
N75:  '/ref4~/rek15~
N76:  '/rec22.c35~
N77:  '{goto}data~
N78:  '/re{end}{down}{right}{right}~/rey3.z35~
N79:  '/cr3.u10~c8~
N80:  '{home}
N81:  '/xmmenu1~
```

Load And Unload Macros

```
N83:  'Recalculate
N84:  '{goto}c22~
N85:  '/xgi80~
```

continued...

...from previous page

```
T26: 'Subroutine dist
T27: '/ppry1.ae33~g
T28: 'ry36.ae38~g
T29: 'ra20.h20~gpq
T30: '/xr

T32: 'Subroutine plotd
T33: '{goto}d36~
T34: '/grgx.{end}{down}~adata~tlotf\c31~
T35: 'ts\c32~tx\c33~ty\c34~ss10~qvq
T36: '/xr

T38: 'Subroutine Ogive
T39: '{goto}ab3~
T40: '/grga.{end}{down}~
T41: 'x{left}{left}{left}.{end}{down}~
T42: 'oss4~tfOgive Based on Frequency Distribution~
T43: 'ts\c32~tx\c34~tyCumulative frequency of occurance~qvq
T44: '{home}/xr

U80: 'Subroutine Gauss2
U81: '/grgxbin~az3.z33~
U82: 'bac3.ac33~
U83: 'otfPlot of Actual vs Expected Value~
U84: 'ts\c32~
U85: 'tx\c34~
U86: 'tyFrequency of occurance~
U87: 'laActual Value~lbExpected Value~
U88: 'ss4~qtlvq
U89: '/xr~

U91: 'Save - saves graph
U92: '/wtc{home}{goto}c8~
U93: '/wwh{window}{goto}I106~
U94: '/g
U96: 'Restart - using \g
U97: '/wwc{home}{goto}c8~/wtb
U98: '/xmmenu4~
```
                Utility Subroutines

continued...

9 Graphic Methods

...from previous page

```
N87:  'Binlim
N88:  '/xnEnter lower bin limit~s14~
N89:  '/xnEnter bin size~s15~
N90:  '/xmmenu1~

Q26:  'Subroutine Summary
Q27:  '/pp1rq19.x22~g
Q28:  'ra1.h20~g
Q29:  '/xr

Q34:  'Sub-MACRO Histogr
Q35:  '/grgxy3.y34~
Q36:  'az3.z34~
Q37:  'otfHistogram of values in data range~
Q38:  'ts\c32~
Q39:  'tx\c34~
Q40:  'tyfrequency of occurance~
Q41:  'ss4~qtbvq
Q42:  '/xr

Q80:  'Subroutine Gauss1
Q81:  '/grgxbin~az3.z33~
Q82:  'bac3.ac33~
Q83:  'otfPlot of Actual vs Expected Value~
Q84:  'ts\c32~
Q85:  'tx\c34~
Q86:  'tyFrequency of occurance~
Q87:  'laActual Value~lbExpected Value~
Q88:  'ss4~qtbvq
Q89:  '/xr~

Q91:  'Plot - Generates tables
Q92:  '/dfy3.y33~s14~s15~~
Q93:  '/rncbin~y3.y33~
Q94:  '/dddata~bin~
Q95:  '{goto}d36~/dfd36~1~1~~
Q96:  '/df.{left}{end}{down}{right}~1~1~~
Q97:  '/c.{left}{end}{down}{right}~{right}~
Q98:  '{home}
Q99:  '/xmmenu1~
```

continued...

...from previous page

```
K106: 'Enter required adjustments to graph
K107: '         Enter (S)ave
K108: '         Enter Graph name
K109: '         Quit Graph mode (q)
K110: '         Press <alt>G to Restart
```

Graph Screen Prompt

Copy The Following Cells over the indicated range

```
AA3:-AA34: (F1) (Z3/$AC$38)*100
AB3:-AB33: (F1) (AB2+AA3)
AB41: 'Gauss
AB42: 'Coef
AB43:-AB73: (F2) (@EXP((-(Y3-$D$8)^2)/(2*($D$9^2))))/$D$9*@SQRT(2*@PI)
AB74: @SUM(AB73..AB43)
AC37: 0
AC38: @SUM(Z3..Z34)
AC39: @SUM(AE43..AE73)
AC3:-AC33: (F1) (AB43/$AB$74)*$F$4
AD3:-AD34: '|
AE3:-AE34: (+) (AA3/2)
AE42: "Chi-Square Coefficient
AE43:-AE73: ((Z3-AC3)^2)/AC3
Y36: 'Low cell contains points below distribution
Y37: 'Pts above High cell
Y38: 'Total Points in distrib.
Y39: 'Chi-Square Coefficient
```

Chi-Sq Calculation Tables

PART III

TWO-WAY ANALYSIS

Two-way analytical techniques are applied to pairs of data files or data files with paired values. These methods evaluate the significance of differences in the two data sets. If the heights of two groups of individuals are recorded, the experimenter might want to know if these individuals are representative of the same population. Given a small set of samples, a manufacturer might want to know if an improvement in a product is indicated. If no shift in mean is indicated, he might want to compare variances to see if a change in producibility has occurred.

Regression techniques work on pairs of data to determine whether two variables are independent of one another. In previous sections, the height of individuals was used as an example in the study of one-way techniques. At the same time, other physical parameters might have been recorded — for example, shoe size. Intuitively, you would expect a close relationship between height and shoe size. The techniques of linear regression allow you to measure how close this relationship is.

Several powerful graphic methods are used for making these comparisons. The **scatter diagram** allows the analyst to look for visual trends when paired data are plotted on Cartesian coordinates. If heights and shoe sizes are plotted, the points fall in a relatively straight line. This is the classic linear, or first order, relationship. If the points align in the shape of a parabola, the relationship is second order. In this case, two variables interact to control a third. Usually, it is desirable to identify both variables and plot them separately. A large number of mathematical models can be identified visually by using scatter plots when simple calculations would not provide the same understanding.

Once you have characterized a population using descriptive statistics, you can represent the population with a Gaussian curve. If you plot two such curves on the same axis, you can evaluate the populations visually. You can measure the probabilities of potential outcomes based on the areas under specific portions of these curves. All of these methods are presented in this section.

Chapter 10 provides a template for these comparisons between data sets. **Linear regression** between pairs of values in a data set can be measured. The coordinates of the regression equation are provided to allow **forecasting** based on the results, which is accomplished with the **Project** command. In addition to regression, a multitude of methods exists for comparison of mean, standard deviations, and variances between data sets. Three have been chosen for their wide applicability and suitability for spreadsheet implementation: the **T test**, the **Z test**, and the **F test**. These methods include a small sample and a large sample comparison for means and one for variances:

- The T test, based on the students' T distribution, is the small sample mean comparison.

- The Z test for means is based on the normal distribution for large samples.

- The F Test on ratios is used to compare variances.

In Chapter 11, templates are provided for scatter plots on X and Y values, plotting of Gaussian curves, and superimposition of least squares lines on scatter plots. Calculations are provided for the intersections of pairs of normal curves. This allows calculation of areas under the curves, which characterizes the probability of various events.

Chapter 12 covers **nonparametric**, or distribution-independent, methods, which are used when nonrandom patterns occur that do not meet the criteria for analysis using standard models.

CHAPTER 10

TWO-WAY ANALYSIS

10 Two-way Analysis

All methods presented so far have been one-way, or single-variable, in nature. These methods provide information about a data set or the population from which it came. The **two-way** techniques in this chapter provide information about how two variables relate. **Regression** techniques look for a dependence of one variable upon another, a common application of which is time dependence. Regression techniques are frequently used to uncover trend information or how a variable performs relative to time.

In addition to regression, there are many techniques for comparing means and deviations of pairs of data sets (to one another or to the population as a whole). The **T test**, based on the students' T distribution, will be used for comparison of means in small samples. The **Z test** is similar in nature but applicable to large samples. As was previously explained, the rule of thumb for what requires large sample analysis is 30 elements. Finally, the **F test** of variance ratio is shown as an evaluation of differences in deviations.

The primary function of this text is to demonstrate how to build statistical models, not to provide the ultimate statistics package. In this next section, several widely useful methods are developed. They are by no means complete, but demonstrate how statistical tests can be integrated into spreadsheet templates. In many cases, adaptations to different formulas can be made with changes as simple as modifying a single equation in a single cell. Other cases will require additions to menus or new templates.

Figure 10.1 is a sample of the output from this template. It is a summary table based on methods described in Chapter 11. Graphic methods are added. In its present state, the template accepts data from a two-column data file, direct entry into the data range, or summary entry into a table. The last is useful for applications where the mean and deviation are already known, making entry of all the data elements unnecessary. Note: an analysis can be performed if data are entered this way. Using statistical tables in conjunction with these tests is necessary. Abbreviated tables are provided in Appendix A, but by using more comprehensive tables, you will gain increased accuracy.

```
              STATISTICAL MODELLING WITH LOTUS 1-2-3
                       Two Way Analysis Template
            File = Exmpl10-1            Operator:Sean Cloake
                         Sample Statistics
                              Standard Deviation
                       Mean  N Weight  N-1 Wght     Count
           Sample X . . . . .  21  11.83215  11.97914      41
           Samply Y . . . . . 29.4  16.56502  16.77080      41

                         Regression Information
                    Y=  -1.1E-14    +     1.4 X
                        Standard err of estimate    1.2E-14
                        Coefficient of correlation        1
                        % of Y variation due to X   100.000

                         T Test   Z Test                       F Test
                        Sml Smpl Lrg Smpl                      ANOVA
           Test Level . .  0.95     0.95    Test Level . .     0.05
           Crit Value . .  1.98     1.96    V1 Value . . .       40
           Deg Freedom. .    74       74    V2 Value . . .       40
           Tails. . . . .     2        2    Crit Value . .     1.69

           Test Value . .-2.60975 -2.64217  Test Value . . 0.510204
           Hypothesis . .     1        1    Hypothesis . .        0
            (Acc=0,Rej=1)                    (Acc=0,Rej=1)
```

Figure 10.1 Two-way Analysis Sample Output

MENUS

There are three menu levels in this template. The autostart menu provides the following options: Test, File, Print, Graph, Cursor, Adjust, and Quit. The Graph option is not used in the template for this chapter, but will be needed for the graphic methods section in Chapter 11. Leaving it in place here allows continuity in the autostart menus. As in all of the templates presented, executing Quit from the primary menu reloads the system menu template for switching between programs. Figure 10.2 summarizes the menu structure.

10 Two-way Analysis

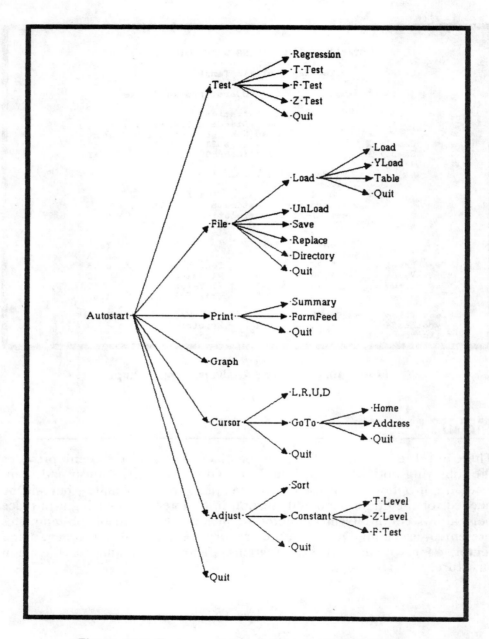

Figure 10.2 Two-way Analysis Template Menu Tree

Two-way Analysis 10

Test

There are four test options in this submenu: **Regression**, **T Test**, **F test**, and **Z test**. The Regression and T test options will cause scratch pad calculations to be made in conjunction with the data ranges in columns C and D. The final calculations are made from these scratch pad calculations and displayed in the summary area. The tables are erased to conserve space in the spreadsheet and avoid slowdown from excessive recalculation.

The F test and Z test calculations can be made from the data in the summary table in cells F31-H34. This table can be loaded directly to allow analysis of data without entry of all the data values, but T test and Regression cannot be performed.

File

The only change in the file menu is the addition of a **Load** submenu, which provides three means of data entry. Load combines a data file into column C. This can be a single file, such as the type generated for the one-way template, or it can be a two-way file. **Load Y** allows a single file to be loaded into column D to provide Y values, which is a convenient way to combine two single files into a double file.

The **Table** entry command accepts statistics and loads them directly into the summary table for use by the Z and F tests. There is an underlying fourth data entry method. The spreadsheet can be halted using **Adjust Halt**. Data can be entered directly into the spreadsheet in columns C and D. Restart, <Alt/R>, performs the summary calculations and transfers control to the primary menu.

The **File Save** option has been modified to save files for two-way analysis.

Cursor and Adjust

The **Cursor** and **Adjust** menus are almost identical to those in the preceding chapter. The **Goto** submenu has been shortened since there is no frequency distribution table to view. The **Recalculate** option is eliminated from the Adjust Menu for the same reason. **Sort** will be used in Chapter 11, and **Constant** has been modified.

10 Two-way Analysis

SPREADSHEET LAYOUT

Figure 10.3 shows the block diagram of the two-way analysis template. This template is very similar to those in previous sections, but is actually simpler. The calculations in this section have fewer intermediate steps. There are only four test subroutines, three file subroutines, and a print routine. In the next chapter, graphics will be added.

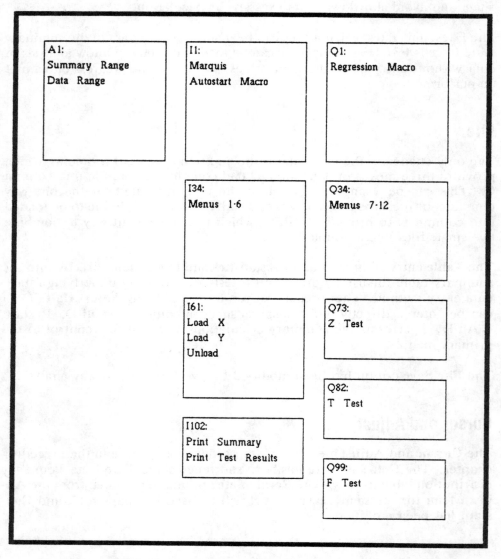

Figure 10.3 Two-way Analysis Template Block Diagram

10 Two-way Analysis

METHODS AND APPLICATIONS

When considering what analytical method to use, you need to consider how many measurements are relevant. For two data sets representing heights of individuals, you might want to determine if there is a difference in height between the groups. It might only be necessary to decide if one group is taller than the other. If both measurements are to be considered, the experiment is referred to as **two-tailed**. If only one measurement is relevant, the experiment is called **single-** or **one-tailed**.

All the equations generated can be used in either single- or two-tailed (one- or two-way) analysis by choosing the critical values carefully. The F, Z, and T test methods are all hypothesis-oriented, which means that two possible outcomes are identified prior to the analysis, and the test statistic is used to select between the two. An example of a single-tailed hypothesis might be as follows:

H_0:   Group A is taller than Group B.

In Figure 10.4, X_0 is the mean of an arbitrary distribution. It could be the first column values in a two-way data file. X is the mean of another distribution, like a second column of values. Two things must be decided to perform a statistical test: the hypothesis and the margin for error.

10 Two-way Analysis

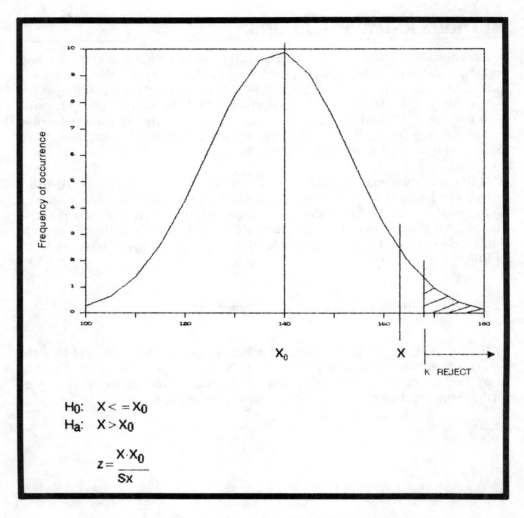

Figure 10.4 Rejection Area for a Single-tailed Experiment

For this figure, the original hypothesis (H_O) is that X is a good estimate of X_O, or, in other words, the two means are the same within statistical limits. The alternate hypothesis (H_a) is as follows:

$H_a: X > X_O$.

All that remains is to select the margin for error. This test is single-tailed so that any value for X is acceptable as long as it is not too high. If you allow a 5% margin for error, then all of that tolerance will be at the top end of the distribution. The rejection area in Figure 10.4 is selected such that the area under that portion of the curve is .05. Remember that the area under a normalized Gaussian curve is 1. If you look at this picture, it should be obvious where the term "single-tailed" originated.

Note that a value K is associated with the edge of the reject area. This "critical value" for the distribution can be found in a statistical table, but it is actually the point on the horizontal scale that cuts the curve so that 5% falls to its right. In a similar manner, a Z value is associated with the mean of the second distribution. In fact, the following is true:

$$Z = \frac{X - X_0}{S_x}$$

If the value of Z is less than the critical value K, then the probability that X is statistically greater than X_0 is smaller than .05. The original hypothesis that X is a good estimate of X_0, then, is accepted, and the means are considered equal. If Z is greater than K, the alternate hypothesis ($H_a: X > X_0$) is accepted, and the means are statistically different.

If the alternate hypothesis were that the two means are not equal, either a high or a low value could reject the hypothesis — hence the term two-tailed. The margin for error is split into two equal areas under the tails of the distribution. The values for K (positive or negative) have a higher associated magnitude because they represent an area of 2.5% on either end. If the Z value exceeds either of the K values, the new alternate hypothesis is accepted; the means are not equal.

With minor adjustments, the T, Z, and F tests presented here can be used for either single or two-tailed analysis. The default is a two-tailed test.

Z Test

The Z test chosen here is based on the Gaussian distribution and is applicable to single- or two-tailed tests for samples of 30 or more. The small sample test based on the T statistic approaches this test as sample size increases. The Z test becomes very insensitive for small sample sizes and, as such, is subject to errors.

10 Two-way Analysis

$$1. \quad z = \frac{\overline{X}-\overline{Y}}{\sqrt{\dfrac{\sigma_x^2}{n_x}+\dfrac{\sigma_y^2}{n_y}}}$$

$$2. \quad T = \frac{\overline{X}-\overline{Y}}{S\sqrt{\dfrac{1}{n_x}+\dfrac{1}{n_Y}}}$$

Where:

$$S = \frac{\sum_{1}^{n_x}(X-\overline{X})^2 + \sum_{1}^{n_Y}(Y-\overline{Y})^2}{n_x+n_y-2}$$

$$3. \quad F = \frac{\sigma_X^2}{\sigma_Y^2}$$

Figure 10.5 Two-way Analysis Equations

Equation #1 in Figure 10.5 is the calculation for the Z statistic. Critical values for this test are found from the table of areas under the normal curve and are independent of sample size. This is advantageous since four values have been included in the spreadsheet, and a table is not required if the test level is set to one of these values. For alpha of .1, .05, .01, or .001, the critical value is automatically provided in the template for single-tailed tests.

If the calculated Z value exceeds the critical value, then it falls in the area under the curve to the right of K in Figure 10.4. For a single-tailed test, this is a rejection of the hypothesis that the means are equal.

T Test

Equation #2 in Figure 10.5 is the small sample test statistic based on students' T. It is very similar to Equation #1, but there is a difference between the two. For large samples, group variance is a good estimator of the variance within the population from which it was selected. In Equation #1, the numerators under the radical are the sample variances. In the small sample case, a more precise estimator of population variance is required. The term S in Equation #2 is an unbiased estimator of the population's variance. The numerator is the number of RMS averages of sample variation from the mean, and the denominator is the number of degrees of freedom. This value is compared to the T statistic for the given accuracy and degrees of freedom.

F Test

The F statistic is based on the ratio of the deviations rather than the difference, as was the case for the two tests on mean. Numerous tables for the F statistic are available with different levels. Most commonly used are the 5% (0.05) and 1% (0.01). As with the T and F statistics, these values still represent the areas under a curve.

Two values are required to decode the F critical value from one of these tables: the number of degrees of freedom from each of the two distributions. If the variables are independent, the degree of freedom is one less than the quantity in each case, which is analogous to the n-1 weighting of the unbiased standard deviation. The F test was chosen because it is accurate over a wide range of sample sizes.

10 Two-way Analysis

Regression

The process of regression fits a straight line (or other curve) to a set of points such that the sum of the deviations from that line is zero. The line has the form of Equation #1 in Figure 10.6. Equations #2 and #3 in Figure 10.6 provide the coefficients for the line described. The applications for such a relationship are limitless. If the data are time and performance, the regression line represents trend performance, which has been applied in manufacturing and marketing departments everywhere. If the two variables are experimental quantities, the relationship can give some measure of whether the two variables are independent. Calculating the value of one variable based on the equation and an arbitrary value of the other allows projection, or forecasting.

$$1. \quad z = \frac{\overline{X} - \overline{Y}}{\sqrt{\frac{\sigma_x^2}{n_x} + \frac{\sigma_y^2}{n_y}}}$$

$$2. \quad T = \frac{\overline{X} - \overline{Y}}{S\sqrt{\frac{1}{n_x} + \frac{1}{n_Y}}}$$

Where:

$$S = \frac{\sum_1^{n_x}(X-\overline{X})^2 + \sum_1^{n_y}(Y-\overline{Y})^2}{n_x + n_y - 2}$$

$$3. \quad F = \frac{\sigma_X^2}{\sigma_Y^2}$$

Figure 10.6 Regression Equations

Given this relationship, three additional pieces of information add significantly to the value of the system. Equation #4 in Figure 10.6 is the standard error in estimation of the points in the line. As with the standard deviation, 65% of the points will fall within one standard error of the line, and 95% will fall within two standard errors. Equation #4 measures the distribution of the accuracy of the individual points.

Equation #5 in Figure 10.6 is the coefficient of determination. When this value is multiplied by 100, it gives the percent of variation in Y that is attributable to X. If Y and X are directly related, then essentially there is only one independent variable, and this value is 100%. The value of the coefficient of correlation in Equation #6 falls between -1 and 1. For a direct relationship, this value is 1. For an inverse relationship, this value is -1. If the value is zero, the variables are unrelated, and the slope of the regression line is zero.

Project

Once the regression equation is known, the **project** test menu option allows a Y value to be projected for an X value entered by the user. The user is prompted to enter an X value, and that value is then plugged into the regression equation. The calculated Y value is displayed. Control is returned to the test menu for continuation or entry of another value.

OUTPUT

The output of this template is shown in Figure 10.1 and is described here for clarity.

Sample Statistics

At the center of the output, near the top, is the Sample Statistics block. The mean, standard deviations, count, and points of intersection for the two samples are calculated for each data group entered. For some problems, typically regressions, this data may not be necessary.

10 Two-way Analysis

The points of intersection will be explained in greater detail in the next chapter, as they relate to graphic methods. A Gaussian curve is projected for each of the two samples in the data set. Two such curves will always intersect in exactly two points. X_1 and X_2 are the values on the first curve associated with the intersection. Y_1 and Y_2 are the values of these points on the second curve.

Regression Information

To the left is the Regression Information block, which contains the equation for the best straight line fit through the given data points. This equation is useful for projections as well as trend indication. The standard error of estimate relates to the line in the same way the standard deviation relates to the mean. 65% of the points in the distribution will fall within one standard error of the line.

Figure 10.7 shows the relationship for standard error. The center straight line is the regression line calculated above. The additional two lines are one standard error from the regression line, and 65.5% of the points in the distribution will fall between these two lines.

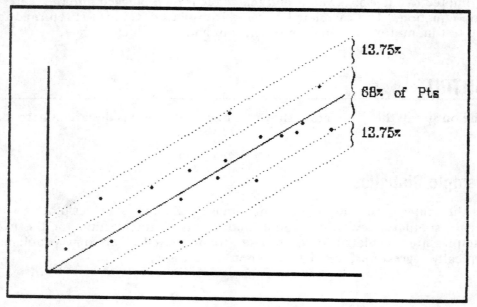

Figure 10.7 Regression Information

The coefficient of correlation is a measure, between zero and one, of how closely the straight line describes the relationship of the points in the distribution. If this value exceeds 0.8, the relationship is probably linear. The percent of Y variation due to X is based on the square of the coefficient of correlation, known as the coefficient of determination. This value, multiplied by 100, is the reported variation in percent.

If height and weight in adult males were compared, the correlation would, expectedly, be fairly high. The coefficient of determination would provide the percentage of variation due to the relationship between height and weight; the remaining variation would be the random size difference in individuals.

Hypothesis Text Results

The remaining three blocks describe the results of the T, Z, and F tests. In each case the test level is fixed but can be changed by editing the spreadsheet to the new value. The critical value is inputted at the time of execution, except in the Z test. Since the value is independent of sample size, no statistical lookup table is required. Two outputs are made from each calculation: the test value and a comparison of this value to the critical value.

COMMANDS USED

Following is the list of commands used in the template created in this chapter:

{edit}{calc}	Converts cell contents to value
@std()	Computes standard deviation
@today	Prints today's date (CPU clock)
@var	Computes variance
/fcce{?}~	Merges file into spreadsheet
/fdb:\~	Changes default directory to B:
/fx{?}~	eXtracts a file or segment
/gs{?}~rq'	Saves a graph

continued...

10 Two-way Analysis

...from previous page

/rncdata~	Creates a range named data
/rnddata~	Deletes a range named data
/xc	Macro gosub
/xg	Macro Goto
/xi	Macro If-Then-Else
/xl	Macro label input
/xn	Macro number input
/xr	Macro return from subroutine
/xm	Macro menu
/wtb	Sets both titles
/wtc	Clears titles

MACRO DESCRIPTIONS

Load Macro

The Load macro in this template has fewer functions than it did in the previous template because fewer calculations are common to all functions in this template. It begins, as before, by requesting and combining a data file, then assigning the data range for the raw data values in the file. Since cell I77 deletes the two data range names, these names must be arbitrarily assigned the first time this template is used. This extra step simplifies operation of the macro. Lotus will not allow you to reassign this name later if it exists, so the step is a worthwile compromise.

```
I73:  'Load Macro
I74:  '{goto}c31~
I75:  '/fcce{?}~
I76:  '{goto}c45~
I77:  '/rnddata~/rnddata1~
I78:  '/rncdata~.{end}{down}~
```

After the X values are assigned to the range named Data, some vital statistics are copied to the summary table. This step is unnecessary if data files are always used. It is possible, however, to load this table directly and perform some of the calculations without entering a whole data file. These few lines of code will allow you to do that later.

Two-way Analysis

```
I79:    '{goto}g31~
I80:    '@avg(data)~{edit}{calc}~
I81:    '{down}@count(data)~{edit}{calc}~
I82:    '{down}@std(data)~{edit}{calc}~
I83:    '{down}@var(data)~{edit}{calc}~
```

The If statement in line 184 has not been used before in this text. It checks the first element in the data file. If the first element is a 2, then there are two columns of data, and the second range may be named (DATA1). The second part of the summary table may also be loaded. If this section of the subroutine is not executed here, it will be executed later.

```
I84:    '/xic31<>2~/xgi92~
I85:    '{goto}d45~
I86:    '/rncdata1~.{end}{down}~
I87:    '{goto}h31~
I88:    '@avg(data1)~{edit}{calc}~
I89:    '{down}@count(data1)~{edit}{calc}~
I90:    '{down}@std(data1)~{edit}{calc}~
I91:    '{down}@var(data1)~{edit}{calc}~
```

Finally, the values from the summary table are copied to the data area. This copying is not necessary except that later a subroutine will be added that allows direct entry of these values for quick calculations.

```
I92:    '/c$g$31~d8~/c$h$31~d9~
I93:    '/c$g$33~e8~/c$h$33~e9~
I94:    '/c$d$8~f8~/c$d$9~f9~
I95:    '{goto}g8~($e$8*@sqrt($g$32/($g$32-1)))~
I96:    '{down}($e$9*@sqrt($h$32/($h$32-1)))~
I97:    '{home}
I98:    '/xr
```

10 Two-way Analysis

Load Y Macro

The Load Y macro is executed from the File Load macro just as the previous Load macro was executed. This macro loads only the Y column. The most normal application of this macro is if two data files created for one-way analysis are to be loaded together for two-way analysis. After the second file is combined, control is transferred to the previously mentioned portion of the Load macro where the second range is named, and the second column of the summary table is calculated.

```
K73:   'LoadY Macro
K74:   '{goto}d31~
K75:   '/fcce{?}~
K76:   '/xgi85~
```

Unload Macro

The Unload macro merely erases all of the data entered by any of the File Loads or test calculations as a means to prepare the spreadsheet for additional test entries.

```
N73:   'Unload Macro
N74:   '/reg31.h34~
N75:   '/rec31.d2048~
N76:   '/rec25.d26~
N77:   '/reg23.g26~
N78:   '/rec13.c13~
N79:   '/ree13.e13~
N80:   '/ref14.f17~
N81:   '/red8.g9~/rec21~
N82:   '/xr
```

Table Subroutine

The Table subroutine allows entry of the summary table on a line-by-line basis. If a Z test or F test is required on a data set where the means, deviations, and counts are known, this routine allows that data to be entered.

```
N84:  'Table Subroutine
N85:  '/xnEnter Avg1 ~g31~
N86:  '/xnEnter Avg2 ~h31~
N87:  '/xnEnter Count1 ~g32~
N88:  '/xnEnter Count2 ~h32~
N89:  '/xnEnted Std Dev1 ~g33~
N90:  '/xnEnter Std Dev2 ~h33~
N91:  '{goto}g34~({up}^2)~
N92:  '/c~{right}~
N93:  '/xgi92~
```

Regression Subroutine

The Regression subroutine calculates the equations of the least squares lines. Once the parameters for the lines are known, this subroutine builds scratch pad tables for calculation of the standard error of estimate and coefficients of correlation and determination. This process can be simplified by Lotus release 2 users, as the regression coefficients can be calculated directly. This macro performs the whole task anyway.

Statements Q2-Q9 generate the sums of X*Y and X^2.

```
Q2:  'Regression Subroutine
Q4:  '{home}{goto}e45~
Q5:  '({left}{left}*{left})~
Q6:  '{right}({left}{left}{left}^2)
Q7:  '{left}/c.{right}~.{left}{end}{down}{right}~
Q8:  '{up}@sum({down}.{end}{down})~
Q9:  '/c~{right}~
```

10 Two-way Analysis

In statements Q9-Q16, the coefficients of the regression line are calculated and stored in the data area. Q10 and Q11 are the numerator and denominator of Equation #2 in Figure 10.6. Q13 puts their ratio in cell E13. Q15 is Equation #3 in Figure 10.6, and it completes the linear equation. Q16 erases the table to make way for the next section.

```
Q10:    '{up}({down}-@count(data)*@avg(data)*@avg(data1))~
Q11:    '{right}({down}-@count(data)*(@avg(data)^2))~
Q12:    '{goto}e13~
Q13:    '(e43/f43)~{edit}{calc}~
Q14:    '{left}{left}
Q15:    '@avg(data1)-@avg(data)*(e13)~{edit}{calc}~
Q16:    '{goto}e43~/re.{right}{end}{down}~
```

Q17-Q20 generate the following four values for the first line in the data table:

```
Q17:    aX + b
Q18:    (Y-Yc)2
Q19:    (Yc-Y)2
Q20:    (Y-Y)Pt2Pt
```

Q21-Q24 copy these values into a table with the sums at the top:

```
Q17:    '{down}{down}({left}{left}*$e$13+$c$13)
Q18:    '{right}({left}{left}-{left})^2~
Q19:    '{right}(({left}{left}-$h$31)^2)
Q20:    '{right}(({left}{left}{left}{left}-$h$31)^2)~
Q21:    '{left}{left}{left}
Q22:    '/c.{right}{right}{right}~.{left}{end}{down}{right}~
Q23:    '{up}(@sum({down}.{end}{down}))~
Q24:    '/c~.{right}{right}{right}~
```

Two-way Analysis 10

Q26-Q28 are the equations for standard error, coefficient of correlation, and percent of Y variation attributable to X. The results are placed in F14-F17 and can be seen in Figure 10.1 as the output values. The remaining statements remove the table upon completion. Remember from previous chapters that {edit}{calc}~ converts the cell contents from formulas to their present values. This combination fixes the result so the tables can be erased. Without it, the results of the calculation would be wiped out when the table range was erased.

```
Q25:    '{home}{goto}f14~
Q26:    '@sqrt(f44/(g32-2))~{edit}{calc}~
Q27:    '{down}@sqrt(g44/h44)~{edit}{calc}~
Q28:    '{down}(g44/h44)*100~/rff3~~{edit}{calc}~
Q29:    '{goto}e44~
Q30:    '/re.{end}{down}{right}{right}{right}~
Q31:    '{home}~
Q32:    '/xr
```

Z Test Subroutine

Cell Q76 is Equation #1 from Figure 10.5. The results are based on the values in the summary table. Q77 compares the Z statistic to the value in D21 and displays 1 if the hypothesis is rejected; it displays 0 if the hypothesis is accepted.

```
Q73:    'Z Test Subroutine
Q75:    '{home}{goto}d25~
Q76:    '(g31-h31)/@sqrt((g34/g32)+(h34/h32)){edit}{calc}~
Q77:    '{down}@if(@abs(d25)>d21,1,0)
Q78:    '{home}
Q79:    '/xr
```

Q84 prompts the user for the critical value of the T statistic, which must be entered from the tables in Appendix A. Q86-Q90 generate the table for calculation of S. Q91 generates the S followed by Q92, the denominator of Equation #2 in Figure 10.5. Q93 moves to C25 and inserts the T statistic for Q94 to compare to the value entered previously. The remaining statements remove the table.

10 Two-way Analysis

T Test Subroutine

```
Q82:    'T Test Subroutine
Q84:    '/xnEnter T critical val ~c21~
Q85:    '{goto}e45~
Q86:    '({left}{left}-$g$31)^2~
Q87:    '{right}({left}{left}-$h$31)^2~
Q88:    '{left}/c.{right}~.{left}{end}{down}{right}~
Q89:    '{up}@sum({down}.{end}{down})~
Q90:    '/c~{right}~
Q91:    '{up}@sqrt(({down}+{down}{right})/(G32+H32-2))~
Q92:    '{right}({left}*@sqrt((1/g32)+(1/h32)))~
Q93:    '{goto}c25~(g31-h31)/f43~{edit}{calc}~
Q94:    '{down}@if(@abs(c25)>c21,1,0)~
Q95:    '{goto}e43~/re.{end}{down}{right}~{goto}a10~
Q96:    '/xr
```

F Test Subroutine

Like the test for the Z statistic, the F test is implemented simply. Q101 positions the cursor so the table can be seen. Q102 prompts for the critical value, which must be entered from the table. Q103 puts the F statistic in the display area, and Q104 provides the hypothesis test result.

```
Q99:    'F Test Subroutine
Q101:   '{home}{pgdn}{up}{up}{up}
Q102:   '/xnEnter F critical value~g23~
Q103:   '{goto}g25~(g34/h34)~
Q104:   '{down}@if(g25>g23,1,0)~
Q105:   '/xr
```

RANGE NAMES

After you enter the spreadsheet in the listing at the end of this chapter, you must declare the following range names before executing the template.

```
/rncdata<return>c45<return>
/rncdata1<return>d45<return>
/rncftest<return>q101<return>
/rncload<return>i74<return>
```

continued...

...from previous page

```
/rncloadY<return>k74<return>
/rncmenu1<return>i35.o3<return>
/rncmenu2<return>i41.n42<return>
/rncmenu3<return>i47.k48<return>
/rncmenu5<return>i60.n61<return>
/rncmenu6<return>i67.l68<return>
/rncmenu8<return>q41.t42<return>
/rncmenu9q<return>47.s48<return>
/rncmenu10<return>q54.u55<return>
/rncmenu12<return>q67.t68<return>
/rncreg<return>q4<return>
/rncsummary<return>i102<return>
/rnctable<return>n85<return>
/rncttest<return>q84<return>
/rncunload<return>n74<return>
/rncztest<return>q75<return>
/rnc\0<return>i24<return>
/rnc\a<return>i24<return>
```

EXAMPLES

Manufacturing Examples

The following table represents cumulative units produced at sequential points in time after starting a production line. Find the equation for a least squares line based on these points. Record the standard error of estimate and the coefficient of correlation for this line. What percent of the variation in units produced is attributable to nonrandom (time-dependent) variation?

Hours	Units
1	91
2	187
5	504
10	938
23	2475
37	3730
49	4895

Table 10.1

10 Two-way Analysis

Using the DATAIN template, store the data from Table 10.1 in a file named exmp10__1. Load the two-way analysis template (TWO__WAY), and input the data file created.

At the primary menu, enter (F)ile (L)oad (L)oadexmp10__1. When the primary menu returns, load TEST REGRESSION. Enter (Q)uit to return to the main menu.

Most of the desired information can be found under the Regression Information heading. The regression equation is as follows:

$$Y = -.059869 + 100.9778 X$$

The standard error of estimate is 79. As with the standard deviation described earlier in this text, this estimate means that 65.5% of the points in this population will fall within 79 units of the line described by the regression equation. The coefficient of correlation is .999272. Since perfect regression is 1.000, the relationship described is very nearly linear. In fact, 99.854 percent of the variation seen is attributable to time, so little random variation is present.

Example #2 relates two runs of a product whose critical parameter is output current. In the final manufacturing stage, the current is measured and recorded. A change in the product has been made, and it is necessary to see if the corresponding change in output is significant. The mean and standard deviation have been recorded for the standard product and new product. Does the difference shown confirm that a positive change in the process has been accomplished?

	Output	Qty	Std Dev
Old Proc	73.000	65	3.8
New Proc	74.700	65	4.7

Since the mean and deviation are already known, loading all the data elements is unnecessary. File Load Table provides the means to enter summarized data directly. In this case, the user enters 73 and 74.7 in response to Avg1 and Avg2; 65 in response to Count1 and Count2; and 3.8 and 4.7 in response to Std Dev 1 and 2.

In this example the following hypotheses will be tested:

H_o: There is no difference in means.
H_a: The means differ.

Since the original hypothesis is testing for difference in means rather than relative magnitude, the two-tailed method is employed. The alternate hypothesis is that the means are different, which is the desired result. The test is accomplished by entering TEST followed by Z-TEST at the main menu.

With a similar method, the deviations can be tested. The hypotheses are as follows:

H_o: The standard deviations differ.
H_a: The standard deviations are the same.

Continuing in the test menu, enter F-TEST, and then quit the menu. At the main menu, the results can be printed, or the cursor can be moved to allow results to be interpreted from the screen. Enter CURSOR DOWN DOWN DOWN. The cursor moves to cell A28, and the Z and F test results are displayed.

The Z test value, -2.26767, exceeds (in magnitude) the critical value, 1.96, so the hypothesis stated previously is rejected. The difference in means is statistically relevant. The F test value, 0.653689, is less than the critical value of 1.53 from the F table in Appendix A. The second hypothesis is rejected; the deviations are not statistically different.

10 Two-way Analysis

SUMMARY

- **Regression** provides information on the relationship of pairs of values, including dependency of one on the other, variability of one with respect to the other, and the degree to which the two correlate.

- The **T test** compares the means of two samples when the samples are small.

- The **F test** compares the ratio of two variances to provide information about the comparative deviations of the two groups.

- The **Z test** compares the means of two groups when sample quantities are large (30).

- In hypothesis testing, two possible outcomes are selected prior to analysis. The **test statistic** is used to decide which of the two to accept.

- In a **single-tailed** test, only one end of the distribution is compared. The hypothesis XY may be rejected for a Y slightly higher than X but will be accepted if Y < X.

- A **two-tailed** test is built on a hypothesis that can be rejected on either end. The hypothesis XY indicates that Y is neither greater than nor less than X within test limits.

- The **Z test** is based on Gaussian distribution with a large sample, so the test statistic is independent of quantity.

- The **critical value** of a test statistic is the ordinate value that intersects the curve such that the area outside the curve equals the desired error probability.

10 Two-way Analysis

EXERCISES

Case Study Four: Manufacturing Example

1. The Idget Corporation is about to release its new, improved Widget. Ten prototype units were built by engineers, followed by 15 pilot units assembled in manufacturing. In the following table, the height of a critical bracket is recorded as measured on these units. Are the pilot units representative of the design as measured on the prototypes? Consider comparison of both means and variances.

   Bracket Height (inches)

Prototypes	Pilot Units
2.15	2.16
2.17	2.16
2.17	2.18
2.20	2.18
2.20	2.19
2.20	2.19
2.21	2.19
2.21	2.19
2.21	2.20
2.24	2.20
	2.20
	2.22
	2.22
	2.23
	2.23

   Table 10.2

2. After the first 100 units are produced, a decision is made to slightly move the height of the critical bracket, with the belief that such a move will allow tighter control in manufacturing. A summary of the results for the first 100 units and the next 35 follows. Was the change in height significant? Was tighter control effected?

10 Two-way Analysis

Analysis of Bracket Move

	Initial	After Change
Mean	2.20	2.22
Deviation	0.015	.011
Samples	100	35

Table 10.3

3. As production increases, records on rate of assembly are kept; they are tabulated below. A large contract has been awarded requiring 100 units per day to be produced. How many operators will be needed at the rate described?

Production Rate

Units/hr	Assemblers
1.5	1
2.7	2
4.33	3
6.1	4
7.05	5

Table 10.4

Answers

1. Enter the data in Table 10.2 into a file named Ex10__1, using the DATAIN template. Load the TWO-WAY template and execute the following:

    <F>ile <L>oad <L>oadX ex10_1

    The mean of both samples is 1.96, and the standard deviations are nearly equal. Since the sample is small, a T test will be used to compare the means.

The hypothesis is that there is no difference in the two groups. The test is executed by entering test mode and T-Test. The critical value of T, from Appendix A, is 2.069 for 23 degrees of freedom (10+15-2) and .05 (1-.95 test level). The test statistic is zero since there is no difference in means. Zero is less than the computed test value, so the hypothesis is accepted. There is no difference.

The F test critical value for samples of 10 and 15 is 2.85. The test value is 1.340236 — much less than the critical value. This hypothesis is also accepted; there is no difference in variances.

Based on these results, the pilot run has performed equivalently to the engineering prototype run.

2. Since summary results are available, creating a data file is unnecessary. Execute the following:

```
<F>ile<L>oad<T>able
2.2                    At Avg1
2.22                   At Avg2
100                    At Count1
35                     At Count2
.015                   At Deviation 1
.011                   At Deviation 2
Test Z-Test
Test F-Test 1.68 <return>
```

Both test values exceed the critical value. Both hypotheses are rejected, so statistically there is a difference in mean and deviation. This means that the bracket has effectively been moved, and in the process the control on variance has been tightened. Remember that in a two-tailed hypothesis test, the difference can be significant in two directions, better or worse. The standard deviation improved, so the general movement was an improvement.

3. There are several logical ways to solve this problem. Cumulative quantities could be calculated as was done in the earlier example. This time, hourly rate will be used. 1000 units/day requires 125 units per eight hours since each assembler works an eight-hour day.

10 Two-way Analysis

The data in the table should be entered into a disk file named ex10__3. This data file is then loaded into the TWO-WAY template. The Regression test provides the following basic equation for production rate:

$$Y = 2.15 + 0.506\ X$$

If the Project option is used, a projection of 63 assemblers is required to produce 125 units/hour. The % variation indicates that 98.197% of the variation in rate is due to staff size.

TEMPLATE LISTING

Following is the complete template listing for this chapter.

```
B2:  '         Two Way Analysis Template
A20: '      Test Level . .
A21: '      Crit Value . .
A22: '      Deg Freedom. .
A23: '      Tails. . . . .
A25: '      Test Value . .
A26: '      Hypothesis . .
A27: '         (Acc=0,Rej=1)
B13: "Y=
B8:  'Sample X . . . . .
B9:  'Samply Y . . . . .
C11: '       Regression Information
C14: 'Standard err of estimate
C15: 'Coefficient of correlation
C16: '% of Y variation due to X
C18: "T Test
C19: 'Sml Smpl
C20: 0.95
C22: @COUNT(C45..C82)+@COUNT(D45..D82)-2
C23: 2
C4:  '       Sample Statistics
D13: ^+
```

continued...

...from previous page

```
D18: "Z Test
D19: 'Lrg Smpl
D20: 0.95
D21: @IF(D20=0.95,1.96,@IF(D20=0.975,2.24,@IF(D20=0.99,2.58,
     @IF(D20=0.999, 3.29, ø))))
D22: (C22)
D23: 2
D6:  '       N Weight
D7:  "Mean
E20: '      Test Level . .
E21: '      V1 Value . . .
E22: '      V2 Value . . .
E23: '      Crit Value . .
E25: '      Test Value . .
E26: '      Hypothesis . .
E27: '         (Acc=0,Rej=1)
E7:  "Dev
F13: 'X
F6:  '     N-1 Weight
F7:  "Mean
G18: "F Test
G19: "ANOVA
G20: 0.05
G21: +G32-1
G22: +H32-1
G7:  "Dev
A29:-H29: '*********

A30: 'Label:
A32: 'File:
A35: 'Variable:
A40: 'Graph:
A45: 'Data:
B31: 'Dmnsion
B32: 'Name
B33: 'Descrip
B34: 'Operator
B35: 'Name
B36: 'Units
B37: 'Hi Limit
B38: 'Lo Limit
```

continued...

10 Two-way Analysis

...from previous page

```
B39:   'Count
B40:   'Title-1
B41:   'Title-2
B42:   'X-title
B43:   'Y-title
B44:   'Flt Lbl
B45:   'Start
C30:   'Data:
F31:   'Avg
F32:   'Count
F33:   'std
F34:   'Var
G30:   ^X
H30:   ^Y

J3:    '***********************************************************
J4:    '***********************************************************
J5:    '***                                                     ***
J6:    '***                                                     ***
J7:    '***         STATISTICAL MODELING WITH LOTUS 1-2-3       ***
J8:    '***                 COMPANION DISKETTE                  ***
J9:    '***                                                     ***
J10:   '***                                                     ***
J11:   '***          A STATISTICAL SOFTWARE PACKAGE             ***
J12:   '***                                                     ***
J13:   '***                                                     ***
J14:   '***                                                     ***
J15:   '***                                                     ***
J16:   '***                                                     ***
J17:   '***                                                     ***
J18:   '***********************************************************
J19:   '***********************************************************
J100:  '       STATISTICAL MODELLING WITH LOTUS 1-2-3
                     Label and Marquis Ranges
```

continued...

Two-way Analysis 10

...from previous page

```
I34: 'Menu1
I35: 'Test
I36: 'Regression  T-Test  Z-Test  F-Test  Quit
I37: '/xmmenu10~
J35: 'File
J36: 'Load  Unload  Save  Replace  Directory  Quit
J37: '/xmmenu2~
K35: 'Print
K36: 'Summary  Distribution  ChiSquare  FormFeed  Quit
K37: '/xmmenu3~
L35: 'Graph
L36: 'Reserved for future use
L37: '/xmmenu1"
M35: 'Cursor
M36: 'Left  Right  Up  Down  GoTo  Quit
M37: '/xmmenu5~
N35: 'Adjust
N36: 'Sort  Constant  Halt  Quit
N37: '/xmmenu6~
O35: 'Quit
O36: 'Retrieve Menu Template
O37: '/fdb:\~
O38: '/frmenu~

I40: 'Menu2 - File
I41: 'Load
I42: 'Load  LoadY  Table  Quit
I43: '/xmmenu8~
J41: 'Unload
J42: 'Remove a file from the template
J43: '/xcunload~
J44: '/xmmenu1~
K41: 'Save
K42: 'Store the current data file
K43: '{goto}c31~
K44: '/fxv{?}~.{pgdn}{end}{down}{right}~
K45: '{home}/xmmenu1~
```

continued...

10 Two-way Analysis

...from previous page

```
L41: 'Replace
L42: 'Save an update to an existing file
L43: '{goto}c31~
L44: '/fxv{?}~.{pgdn}{end}{down}{right}~r
L45: '{home}/xmmenu1~
M41: 'Directory
M42: 'Change the directory
M43: '/fd{?}~
M44: '/xmmenu1~
N41: 'Quit
N42: 'Return to the Primary Menu
N43: '/xmmenu1~

I46: 'Menu3 - Print
I47: 'Summary
I48: 'Print the summary report
I49: '/xcsummary~
I50: '/xmmenu3~
J47: 'FormFeed
J48: 'Advance page to top of form
J49: '/pppq
J50: '/xmmenu3~
K47: 'Quit
K48: 'Return to main menu
K49: '/xmmenu1~

I59: 'Menu5 - Cursor
I60: 'Left
I61: 'Move cursor left five columns
I62: '{left}{left}{left}{left}{left}
I63: '/xmmenu5~
J60: 'Right
J61: 'Move cursor right five columns
J62: '{right}{right}{right}{right}{right}
J63: '/xmmenu5~
K60: 'Up
K61: 'Move cursor up one page
K62: '{up}{up}{up}{up}{up}{up}{up}{up}{up}
K63: '/xmmenu5~
```

continued...

Two-way Analysis 10

...from previous page

```
L60: 'Down
L61: 'Move cursor down one page
L62: '{down}{down}{down}{down}{down}{down}{down}{down}{down}
L63: '/xmmenu5~
M60: 'GoTo
M61: 'Home   Address
M62: '/xmmenu9~
N60: 'Quit
N61: 'return to main menu
N62: '{home}
N63: '/xmmenu1~

I66: 'Menu6 - Adjust
I67: 'Sort
I68: 'Reserved for future application
I69: '/xmmenu1~
J67: 'Constant
J68: 'Low-Bin  Bin-Size  Chi-Level
J69: '{goto}q12~
J70: '/xmmenu12~
K67: 'Halt
K68: 'Halt the MACRO to make changes
K69: '/wtc{home}
K70: '/xq
L67: 'Quit
L68: 'Return to primary menu
L69: '/xmmenu1~


:    'Menu8 - Load options
Q41: 'Load
Q42: 'Load X column or X,Y columns
Q43: '/xcload~
Q44: '/xmmenu1~
R41: 'LoadY
R42: 'Load Y column only
R43: '/xcloady~
R44: '/xmmenu1~
S41: 'Table
S42: 'Load Data into Calculation Table
```

continued...

10 Two-way Analysis

...from previous page

```
S43:  '/xctable~
S44:  '/xmmenu1~
T41:  'Quit
T42:  'Return to Main Menu
T43:  '/xmmenu1~

Q46:  'Menu9 - GoTo Options for Cursor
Q47:  'Home
Q48:  'Home Cursor
Q49:  '{home}
Q50:  '/xmmenu1~
R47:  'Address
R48:  'GoTo a specific cell
R49:  '{goto}{?}~
R50:  '/xmmenu9~
S47:  'Quit
S48:  'Return to main menu
S49:  '/xmmenu1~

Q53:  'Menu10 - Test Options
Q54:  'Regression
Q55:  'Compute Regression between X and Y
Q56:  '/xcreg~
Q57:  '/xmmenu10~
R54:  'T-Test
R55:  'Perform Small Sample test on means of X and Y
R56:  '/xcttest~
R57:  '/xmmenu10~
S54:  'F-Test
S55:  'Perform F test on ratio of variances
S56:  '/xcftest~
S57:  '/xmmenu10~
T54:  'Z-Test
T55:  'Perform Large Sample test on means of X and Y
T56:  '/xcztest~
T57:  '/xmmenu10~
U54:  'Quit
U55:  'Return to main menu
U56:  '/xmmenu1~
```

continued...

...from previous page

```
Q66:  'Menu12 - Change Constants
Q67:  'T Test level
Q68:  'Change the T test level
Q69:  '/xnEnter new T test level ~c20~
Q70:  '/xmmenu12~
R67:  'Z Test level
R68:  'Change the Z Test level
R69:  '/xnEnter new Z test level ~d20~
R70:  '/xmmenu12~
S67:  'F Test level
S68:  'Change the F test level
S69:  '/xnEnter new F test level ~g20~
S70:  '/xmmenu12~
T67:  'Quit
T68:  'Return to main menu
T69:  '{home}
T70:  '/xmmenu6~
```

Menu Macros

```
I23:  'start
I24:  '{home}{goto}i1~{goto}k15~/cI32~~
I25:  '{home}/rek15~
I26:  '/xmmenu1~
```

Autostart MACRO

```
I101: 'Print Summary Subroutine
I102: '/ppcara29.h29~g
I103: 'lri100.p100~g
I104: 'lra1.h29~gq
I105: '/xr
```

Summary Print MACRO

```
I73:  'Load MACRO
I74:  '{goto}c31~
I75:  '/fcce{?}~
I76:  '{goto}c45~
I77:  '/rnddata~/rnddata1~
I78:  '/rncdata~.{end}{down}~
I79:  '{goto}g31~
```

continued...

10 Two-way Analysis

...from previous page

```
I80: '@avg(data)~{edit}{calc}~
I81: '{down}@count(data)~{edit}{calc}~
I82: '{down}@std(data)~{edit}{calc}~
I83: '{down}@var(data)~{edit}{calc}~
I84: '/xic31<>2~/xgi92~
I85: '{goto}d45~
I86: '/rncdata1~.{end}{down}~
I87: '{goto}h31~
I88: '@avg(data1)~{edit}{calc}~
I89: '{down}@count(data1)~{edit}{calc}~
I90: '{down}@std(data1)~{edit}{calc}~
I91: '{down}@var(data1)~{edit}{calc}~
I92: '/c$g$31~d8~/c$h$31~d9~
I93: '/c$g$33~e8~/c$h$33~e9~
I94: '/c$d$8~f8~/c$d$9~f9~
I95: '{goto}g8~($e$8*@sqrt($g$32/($g$32-1)))~
I96: '{down}($e$9*@sqrt($h$32/($h$32-1)))~
I97: '{home}
I98: '/xr
```

Load Macro

```
K73: 'LoadY MACRO  K74: '{goto}d31~
K75: '/fcce{?}~
K76: '/xgi85~
N73: 'Unload MACRO
N74: '/reg31.h34~
N75: '/rec31.d2048~
N76: '/rec25.d26~
N77: '/reg23.g26~
N78: '/rec13.c13~
N79: '/ree13.e13~
N80: '/ref14.f17~
N81: '/red8.g9~/rec21~
N82: '/xr
```

LoadY MACRO

continued...

Two-way Analysis 10

...from previous page

```
N84: 'Table Subroutine
N85: '/xnEnter Avg1 ~g31~
N86: '/xnEnter Avg2 ~h31~
N87: '/xnEnter Count1 ~g32~
N88: '/xnEnter Count2 ~h32~
N89: '/xnEnted Std Dev1 ~g33~
N90: '/xnEnter Std Dev2 ~h33~
N91: '{goto}g34~({up}^2)~
N92: '/c~{right}~
N93: '/xgi92~
```
              Enter Summary Table Subroutine


```
Q2:  'Regression Subroutine
Q4:  '{home}{goto}e45~
Q5:  '({left}{left}*{left})~
Q6:  '{right}({left}{left}{left})^2)
Q7:  '{left}/c.{right}~.{left}{end}{down}{right}~
Q8:  '{up}@sum({down}.{end}{down})~
Q9:  '/c~{right}~
Q10: '{up}({down}-@count(data)*@avg(data)*@avg(data1))~
Q11: '{right}({down}-@count(data)*(@avg(data)^2))~
Q12: '{goto}e13~
Q13: '(e43/f43)~{edit}{calc}~
Q14: '{left}{left}
Q15: '@avg(data1)-@avg(data)*(e13)~{edit}{calc}~
Q16: '{goto}e43~/re.{right}{end}{down}~
Q17: '{down}{down}({left}{left}*$e$13+$c$13)
Q18: '{right}({left}{left}-{left})^2~
Q19: '{right}(({left}{left}-$h$31)^2)
Q20: '{right}(({left}{left}{left}{left}-$h$31)^2)~
Q21: '{left}{left}{left}
Q22: '/c.{right}{right}{right}~.{left}{end}{down}{right}~
Q23: '{up}(@sum({down}.{end}{down}))~
Q24: '/c~.{right}{right}{right}~
Q25: '{home}{goto}f14~
Q26: '@sqrt(f44/(g32-2))~{edit}{calc}~
Q27: '{down}@sqrt(g44/h44)~{edit}{calc}~
Q28: '{down}(g44/h44)*100~/rff3~~{edit}{calc}~
Q29: '{goto}e44~
```

continued...

10 Two-way Analysis

...from previous page

Q30: '/re.{end}{down}{right}{right}{right}~
Q31: '{home}~
Q32: '/xr
Regression Subroutine

Q73: 'Z Test Subroutine
Q75: '{home}{goto}d25~
Q76: '(g31-h31)/@sqrt((g34/g32)+(h34/h32)){edit}{calc}~
Q77: '{down}@if(@abs(d25)>d21,1,0)
Q78: '{home}
Q79: '/xr
Z Test Subroutine

Q82: 'T Test Subroutine
Q84: '/xnEnter T critical val ~c21~
Q85: '{goto}e45~
Q86: '({left}{left}-g31)^2~
Q87: '{right}({left}{left}-h31)^2~
Q88: '{left}/c.{right}~.{left}{end}{down}{right}~
Q89: '{up}@sum({down}.{end}{down})~
Q90: '/c~{right}~
Q91: '{up}@sqrt(({down}+{down}{right})/(G32+H32-2))~
Q92: '{right}({left}*@sqrt((1/g32)+(1/h32)))~
Q93: '{goto}c25~(g31-h31)/f43~{edit}{calc}~
Q94: '{down}@if(@abs(c25)>c21,1,0)~
Q95: '{goto}e43~/re.{end}{down}{right}~{goto}a10~
Q96: '/xr
T Test Subroutine

Q99: 'F Test Subroutine
Q101: '{home}{pgdn}{up}{up}{up}
Q102: '/xnEnter F critical value~g23~
Q103: '{goto}g25~(g34/h34)~
Q104: '{down}@if(g25>g23,1,0)~
Q105: '/xr
F Test Subroutine

CHAPTER 11

TWO-WAY GRAPHIC METHODS

11 Two-way Graphic Methods

The graphic methods presented here augment the two-way methods in Chapter 10. The previous measures provide either a quantitative relationship between the data sets or a go/no-go indication of the significance of the difference. If the magnitude of the difference indicator exceeds a standardized value, it is considered significant. Graphic methods allow the user to look for trends in the data that may not be apparent from calculations, compare populations from which samples are taken, and estimate probabilities of occurrence of certain values.

An **X-Y** scatter plot—or its inverse, the **Y-X** scatter plot—presents visually the relationship described by the coeffecint of correlation of the regression equation. If the coefficient is near 1, the points fall closely on either side of a straight line described by the regression equation. Certain other mathematical patterns can be identified by grouping data visually. For example, a second-order relationship would cause points to describe a parabola. Power and exponential curves also have readily identifiable characteristic curves. These curves can be fit mathematically, but that discussion is not within the scope of this book.

Data presented in scatter plots reflect random variation if measurement techniques are not accurate. Also, certain data types are subject to seasonal variations that can affect regression calculations and plots. The sales of a toy company, plotted monthly, would be subject to peaks every June and December as a result of summer and Christmas boosts. **Curve-smoothing** is a technique in which each point is replaced by the average of itself and its neighbors. The overall effect is to distribute the measurement errors or cyclic variations, both of which are short-term, over a group of cells. Long-term variations can then be seen more easily.

The **two-way Gaussian** plot displays the probability density function of both sample populations on the same axis. These two curves always intersect in exactly two points, which are calculated and displayed in the summary report. The areas under different portions of the curves define the probabilty of potential outcomes, since the total area under each curve is unity. These areas can be found easily from the mean, deviation, and distance of the point from the mean.

Two-way Graphic Methods 11

Figure 11.1 is an amended version of the output from this template. It has been divided into two sections that can be printed separately. The top section now contains the points of intersecton of two Gaussian curves with means and deviations described by the values in the table. The four Z values are the standardized normal random variables (NRVs) for the two points of intersecton and the two data sets. Use of these variables will be explained later.

```
              STATISTICAL MODELLING WITH LOTUS 1-2-3
                     Two Way Analysis Template

         File: Examp11_1           Operator: Patty

                    Sample Statistics
                              Standard Deviation
                      Mean    N Weight   N-1 Wght      Count
         Sample X . . .10.08333  9.304643  9.718383      12
         Sample Y . . .20.1585  18.32276  19.13750       12
         Points of Intersection -8.25042  21.41503

              Regression Information              Standardized NRV
         Y= 0.310632      +     1.968383 X         X1=   -1.89
         Standard err of estimate. .  0.580330     X2=    1.17
         Coefficient of correlation.  0.999581     Y1=   -1.48
         % of Y variation due to X .  99.916       Y2=    0.07
*****************************************************************
                     T Test    Z Test                    F Test
                    Sml Smpl  Lrg Smpl                   ANOVA
         Test Level . .  0.95     0.95     Test Level . .  0.05
         Crit Value . .           1.96     V1 Value . . .    11
         Deg Freedom. .    22       22     V2 Value . . .    11
         Tails. . . . .     2        2     Crit Value . .
         Test Value . .                    Test Value . .
         Hypothesis . .                    Hypothesis . .
         (Acc=0,Rej=1)                     (Acc=0,Rej=1)
```

Figure 11.1 Graphic Template Output

MENUS

The menu in Figure 11.2 is similar to that in Figure 10.2. Elements in italics are the additions for this section. The Print menu has been modified to reflect the split in the summary table. A **Graph** menu is added with a **Create** submenu to call the various subroutines to create graphs. A **Sort** submenu is added to allow data to be sorted on either X or Y or returned to their original values.

11 Two-way Graphic Methods

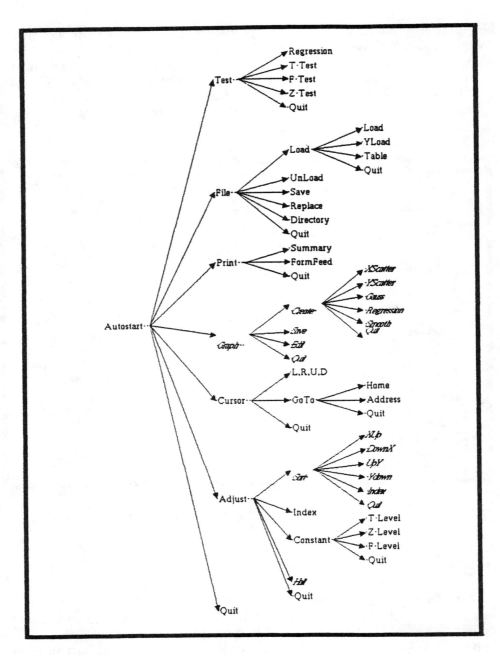

Figure 11.2 Two-way Graphic Template Menu Tree

11 Two-way Graphic Methods

METHODS AND APPLICATIONS

The material presented in this graphics template has been merged with the two-way analysis template presented in Chapter 10. Figure 11.3 shows the modified template. The following sections describe these added sections and their applications.

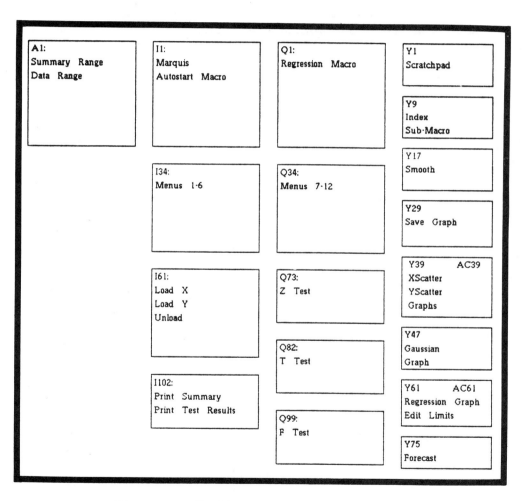

Figure 11.3 Modified Template Block Diagram

11 Two-way Graphic Methods

X-Y and Y-X Scatter Plots and Regression

If the data in columns X and Y are closely related — as with a correlation coefficient near 1 — there is no difference between this scatter plot and the line plot of Chapter 9. The value of a scatter plot surfaces when there is no linear correlation.

The data in the following table have been plotted using this X-Y plot option. The result is shown in Figure 11.4. It is obvious that the data are not linearly related, but, clearly, the variation is not random. This data is, in fact, related by a second-order relationship. This is an example of the value of visual interpretation of data.

X	Y
-5	25
-4	16
-3	9
-2	4
-1	1
0	0
1	1
2	4
3	9
4	16
5	25

Table 11.1 Data for Figure 11.4

Two-way Graphic Methods 11

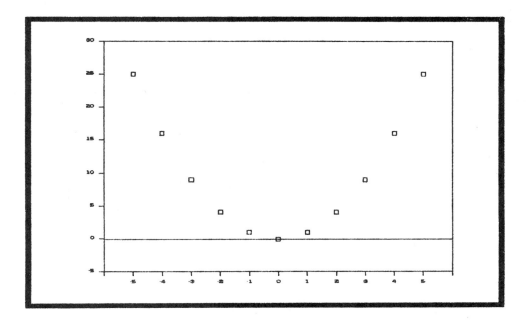

Figure 11.4 X-Y Plot Example

The Y-X plot option allows the analyst to reverse the axis and replot the data. Frequently, the relationship is a function of Y — not X — and will show up more clearly after this reversal.

The regression plot in Figure 11.5 is similar to the X-Y plot described previously, but a straight line has been added. This line is generated by extrapolating new values for Y from the regression equation and X values provided. If correlation is high, the analyst can evaluate how closely the points are grouped. Remember, 65% of the points will fall within 1 standard error of the line. If the correlation is low, the randomness of the points can be checked.

11 Two-way Graphic Methods

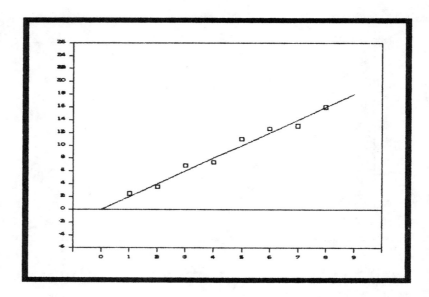

Figure 11.5 Regression Plot Example

The most common use of a regression line superimposed on a set of points is to report data trends. If the slope of the regression line is negative, the data shows a downward trend. A positive slope indicates an increasing trend.

All techniques described in this section are suited to paired data sets. In these cases, data are usually taken as pairs and recorded together. This is consistent with the height and shoe size example from the last chapter in which both values were recorded for a group of individuals.

Smoothing

When measurement methods are poor, random error is introduced and the accuracy of individual measurements is lessened. This problem tends to affect variance more than mean value, but it is not a measure of variation in the data set. It is representitive only of the variation in the measurement. **Smoothing** allows an analyst to check for relationships between X and Y with error redistributed. If each point is replaced by the average of itself and its neighbors, these errors are minimized.

On any point, the errors are distributed around the real value with equal probability of an error being high or low. The central limit theorem states that as the number of points in the average increases, the averaged value approaches the real value. Generally, an odd number of samples is chosen — between three and eleven. For the purposes of this template, three-point smoothing has been used. The smoothed curve, then, approaches the real curve as a limit.

Seasonal variations can also be smoothed using this technique. Many businesses have short-term (or seasonal) variations. Toy sales for Christmas are always high. If a toy company plotted a trend chart for its first year of business, starting in January, they might be misled. A simple trend chart would indicate that business is skyrocketing when, in fact, only seasonal variation is taking place. If the December point were averaged with January and November, the seasonal effect would be significantly reduced; however, the November and January readings are increased as December is averaged into them.

Gaussian

Data for a **Gaussian** plot need not come in pairs. In this case, two normal curves are generated on a Cartesian axis. The mean and deviation of these curves are taken from the two data sets in X and Y. In effect, a best estimate Gaussian curve is generated for each data set, and both of these curves are plotted on the axis.

Several interesting relationships can be identified if you know the points where these two curves intersect. Following is a derivation for a set of equations to identify these points. Since the Gaussian equation is second-order, the two curves intersect in exactly two points, as mentioned earlier. It is a good idea to use a scratch pad to convince yourself that this is true because it is not always obvious.

To keep the equations simple, some substitutions are made at the start:

$$\text{LET:} \quad (1) \quad \begin{aligned} t &= \mu_1 = G31 \\ u &= \sigma_1 = F8 \\ v &= \mu_2 = H31 \\ w &= \sigma_2 = F9 \end{aligned}$$

Equation 11.1

11 Two-way Graphic Methods

This substitution allows the use of English letters rather than Greek symbols in the following equations. Equations for the two Gaussian curves are generated and set equal to each other. The roots of the resultant equation will provide the two common points.

$$f(x1) = \frac{e^{\frac{-(x-t)^2}{2u^2}}}{u\sqrt{2\pi}} = \frac{e^{\frac{-(x-v)^2}{2w^2}}}{w\sqrt{2\pi}} = f(x2)$$

Equation 11.2

$$e^{\frac{-(x-t)^2}{2u^2}} = \frac{u}{w} e^{\frac{-(x-v)^2}{2w^2}}$$

Equation 11.3

Take the Naperian log of both sides:

$$-(x - t)^2/2u^2 = \ln(u/w) - (x - v)^2/2w^2$$

Equation 11.4

Finally, reduce algebraically:

$$-w^2(x - t)^2 = 2u^2w^2\ln(u/w) - u^2(x - v)^2$$

Equation 11.5

$$-w^2x^2 + 2w^2tx - w^2t2 = 2u^2w^2\ln(u/w) - u^2x^2 + 2u^2vx - u^2v^2$$

Equation 11.6

318

Two-way Graphic Methods 11

$$(u^2 - w^2)x^2 + 2(w^2t - u^2v)x + (u^2v^2 - w^2t^2 - 2u^2w^2\ln(u/w)) = 0$$

Equation 11.7

This equation is now in the following form:

$$aX^2 + bX + c = 0$$

If:

$$a = u^2 - w^2$$

Equation 11.8

$$b = 2(w^2t - u^2v)$$

Equation 11.9

$$c = (u^2v^2 - w^2t^2 - 2u^2w^2\ln(u/w))$$

Equation 11.10

Use the quadratic equation:

$$x = -b/2a +/- (\sqrt{b^2 - 4ac}/2a)$$

Equation 11.11

11 Two-way Graphic Methods

Equations can be developed for the two terms in this equation. Notice that as soon as the two means and deviations are known, the points of intersection can be calculated automatically by including these equations in cells on the spreadsheet.

$$S_1 = (u^2v^2 - w^2t^2)/(n^2 - w^2)$$

Equation 11.12

$$S_2 = \frac{\sqrt{(2w^2t - 2u^2v)^2 - 4(u^2 - w^2)(u^2v^2 - w^2t^2 - 2u^2w^2\ln(u/w))}}{2(u^2 - w^2)}$$

Equation 11.13

The two roots, or X values, are

$$X1 = S_1 - S_2$$
$$X2 = S_1 + S_2$$

The equations for X1 and X2 are entered into the spreadsheet in the regression subroutine. Substitute the cell locations from equations in Equation 11.1 for t, u, v, and w. The actual equations were spread over several cells to make debugging the macros easier.

Figure 11.6 is typical of plots generated by this function. A few pitfalls should be avoided. The curves are based on the unbiased (n-1 weighted) deviation, but they still do not tend to approach reality until 30 or more points are used. Furthermore, a few conditions exist in which the points of intersection are very far from the mean. One example is a case with nearly equal deviations and differing means. In some cases, the result of the calculation for the roots may be undefined. Such distributions don't tend to provide significant information graphically.

Two-way Graphic Methods **11**

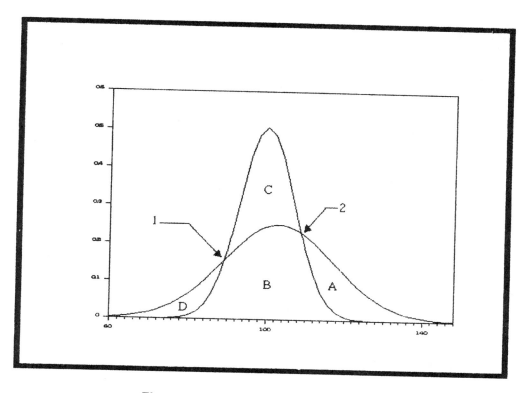

Figure 11.6 Gaussian Intersection Plot

There are four areas in the graph in Figure 11.6, marked A, B, C, and D. Keeping in mind that the area under each of the two curves is unity, you can calculate some probabilities. If the two curves represent outputs from two possible production processes, the areas each have a significance. Area A is the probability that a point taken from the right-hand sample indicates a better process than the left. Area D is the counter-case that this distribution is worse. The fact that one process can be better or worse at the same time is a result of one distribution having a wider variance than the other. Area B is common to both distributions. Area C is the probability that the performance of process 1 will be worse than that of 2.

It is a simple matter to find any of these exact areas from the table of areas under the normal curve and the four values in the summary table. X1 and X2 are the standardized normal random variates for the two intersects relative to the mean and deviation for distribution 1. Y1 and Y2 relate to distribution 2.

11 Two-way Graphic Methods

OUTPUT

The summary table has been made into two tables that can be printed independently, which allows creation of output reports including only the summary results and regression coupled with one or more graphs. If you want, the Z, F, and T Test results can be added. To this output has been added the points of intersection of the Gaussian curves and the four standardized normal random variates in addition to the five graphs detailed previously.

To summarize, the outputs of this template include the following:

- X-Y Scatter Graph
- Y-X Scatter Graph
- Plot of two Gaussian curves
- Scatter plot with regression line
- Three-point smoothed graph
- Coordinates for intersection of two curves (Gaussian)
- Normalized values for coordinates

COMMANDS USED

Following is the list of commands used in the template created in this chapter:

Command	Description
{edit}{calc}	Converts cell contents to value
@std()	Computes standard deviation
@today()	Prints today's date (CPU clock)
@var()	Computes variance
/fcce{?}~	Merges file into spreadsheet
/fdb:\~	Changes default directory to B:
/fx{?}~	eXtracts a file or segment
/gs{?}~	rq Saves a graph
/rncdata~	Creates a range named data
/rnddata~	Deletes a range named data
/xc	Macro gosub
/xg	Macro Goto
/xi	Macro If-Then-Else
/xl	Macro label input

continued...

...from previous page

/xn   Macro number input
/xr   Macro return from subroutine
/xm   Macro menu
/wtb   Sets both titles
/wtc   Clears titles

MACRO DESCRIPTIONS

Unload Macro

The following Unload Macro has been expanded to include the ranges added for this section.

```
N73:    'Unload MACRO
N74:    '/reg31.h34~
N75:    '/rec31.d2048~
N76:    '/rec25.d26~/ree40.h2048~
N77:    '/reg23.g26~
N78:    '/reb13~/rec3~
N79:    '/red13~/ref3~
N80:    '/ree14.e16~
N81:    '/red8.g9~/rec21~
N82:    '/xr
```

Print Summary Subroutine

The Print Summary subroutine has been shortened as compared to the one shown in Chapter 10. The commands that were removed are in the Test Results subroutine.

```
I101:   'Print Summary Subroutine
I102:   '/ppcara29.h29~g
I103:   'ri100.p100~g
I104:   'ra1.h16~gra29.h29~gq
I105:   '/xr
```

Test Results Subroutine

The Test Results subroutine appends the test results to the summary report.

```
M101:   'Print test results
M102:   '/ppcara18.h23~g
M103:   'ra25.h27~gra29.h29~gq
M104:   '/xr
```

Scratchpad Subroutine

The scratchpad area calculates the intersections of the two Gaussian curves as soon as data are entered. Z3, Z4, and Z5 are the three coordinates of the equation

$$a X 2 + b X + c = 0$$

```
Y1:   'Scratchpad
Y3:   "a=
Y4:   "b=
Y5:   "c=
Z3:   ((F8^2)-(F9^2))
Z4:   2*(G31*(F9^2)-H31*(F8^2))
Z5:   (F8*H31)^2-(F9*G31)^2-2*(F8*F9)^2*@LN(F8/F9)
```

AB3 and AB4 are the two terms of the quadratic equation, which provides the two intersection points:

```
AA3:  "S1=
AA4:  "S2=
AB3:  -Z4/(2*Z3)
AB4:  (@SQRT((Z4^2)-4*Z3*Z5))/(2*Z3)
```

Smooth Subroutine

Line Y20 contains the average of the first three values in the data range. Y21 copies this value, generating a range of three-point smoothed values next to Y; that is, each value is the average of three corresponding values in Y.

```
Y17:    'Smooth Subroutine
Y19:    '/ref45.f2048~{goto}f46~
Y20:    '@sum(d45.d47)/3~
Y21:    '/c~.{left}{left}{end}{down}{right}{right}~
```

The remainder of the subroutine generates the smooth graph. The graph ranges begin on the second line and end in the next-to-last line in the X range, eliminating the top and bottom values. The end values in the smoothed range are the average of three numbers, one of which is a zero from the blanks above and below the data range.

```
Y22:    '/grgxdata~x{up}{up}~adata1~a{up}{up}~
Y23:    'b.{end}{down}{up}{up}~oss4~
Y24:    'tfRaw Data vs Three Point Smoothed~
Y25:    'ts\c40~tx\c42~ty\c43~qvq
Y26:    '/ree45.h2048~{home}
Y27:    '/ree45.h2048~/xr~
```

Xscat and Yscat Subroutines

Xscat and Yscat generate the scatter plot graphs that are called from menu 7:

```
Y39:    'Xscat
Y40:    '/grgxdata~adata1~
Y41:    'oss4~
Y42:    'tfX-Y Scatter Plot~
Y43:    'ts\c40~tx\c44~ty\d44~qvq
Y44:    '/ree45.h2048~{home}
Y45:    '/xr~
```

11 Two-way Graphic Methods

```
AC39:   'Yscat
AC40:   '/grgxdata1~adata~
AC41:   'oss4~
AC42:   'tfY-X Scatter Plot~
AC43:   'ts\c40~tx\d44~ty\c44~qvq
AC44:   '/ree45.h2048~{home}
AC45:   '/xr~
```

Gauss Subroutine

The Gauss subroutine generates a table and uses it to create the graph with two Gaussian curves. Line Y49 generates the X values for the range. It fills a column starting with the lowest of the two minimum values provided in the data range, ending at the highest of the maximums, and with a cell size calculated to provide 100 cells in the column.

```
Y47:   'Gauss
Y48:   '{goto}e45~
Y49:   '/dfe45.e155~g35~g37~g36~
```

Lines Y50-51 are the equations for the Gaussian Probability Density Functions based on the two means and deviations. These equations are copied into two columns next to the X values from Y49.

```
Y50:   (F2)'{right}(@EXP((-({left}-
       $d$8)^2)/(2*($f$8^2))))/$f$8*@SQRT(2*@PI)~
Y51:   (F2)'{right}(@EXP((-({left}{left}-
       $d$9)^2)/(2*($f$9^2))))/$f$9*@SQRT(2*@PI)~
Y52:   '/cf45.g45~f45.f145~
```

The remaining statements generate the required graph:

```
Y53:   '/grgx{left}{left}.{end}{down}~
       a{left}.{end}{down}~b.{end}{down}
Y54:   'otfIntersection of Two Populations~ts\c40~
Y55:   'tx\c35~tyMagnitude of PDF~q
Y56:   'oss20~falblqqvq
Y57:   '/ree45.h2048~{home}
Y58:   '/xr~
```

Two-way Graphic Methods 11

Regression Graph Subroutine

The Regression subroutine is similar to the Gaussian routine above. Lines Y63-64 create a column of Y values related to the given X values by the regression equation calculated for the data set. The remainder of the subroutine generates the graph.

```
Y61:  'Regression graph
Y62:  '{goto}e45~
Y63:  '({left}{left}*$d$13+$b$13)~
Y64:  '/c~.{left}{end}{down}{right}~
Y65:  '/grgx{left}{left}.{end}{down}~
Y66:  'a{left}.{end}{down}~
Y67:  'b.{end}{down}~
Y68:  'ofasblqss5~
Y69:  'tfData Points With Regression Line~
Y70:  'ts\c40~tx\c42~ty\c43~qvq
Y71:  '/ree45.h2048~{home}
Y72:  '/xr~
```

RANGE NAMES

After entering the spreadsheet in the listing at the end of this chapter, you must declare the following range names prior to executing the template.

```
/rncdata<return>c45<return>
/rncdata1<return>d45<return>
/rncftest<return>q101<return>
/rncload<return>i74<return>
/rncloadY<return>k74<return>
/rncmenu1<return>i35.o36<return>
/rncmenu2<return>i41.n42<return>
/rncmenu3<return>i47.k48<return>
/rncmenu5<return>i60.n61<return>
/rncmenu6<return>i67.l68<return>
/rncmenu8<return>q41.t42<return>
/rncmenu9<return>q47.s48<return>
/rncmenu10<return>q54.u55<return>
/rncmenu12<return>q67.t68<return>
/rncreg<return>q4<return>
/rncsummary<return>i102<return>
```

continued...

11 Two-way Graphic Methods

...from previous page

```
/rnctable<return>n85<return>
/rncttest<return>q84<return>
/rncunload<return>n74<return>
/rncztest<return>q75<return>
/rnc\0<return>i24<return>
/rnc\a<return>i24<return>
```

SUMMARY

- Two Gaussian curves always intersect in exactly two places.

- Since areas under Gaussian curves are normalized to unity, each area is, in itself, a probability.

EXERCISES

1. Using the data file created in the first exercise in Chapter 10, plot the two Gaussian curves associated with those distributions.

2. Plot a regression graph for the data in Exercise 10.3.

3. Enter the following data into a file named "ex11__1." This data represents two manufacturing processes. If a sample is pulled from the second process, what is the probability that it will be better than a sample from the first process? (**Note:** data are in histogram form. The number 3 in the Proc1 column indicates that you should put three 110's in the left data column.)

	Proc1	Proc 2
100	1	0
110	3	0
120	8	1
130	17	4
140	19	9
150	16	16

continued...

...from previous page

160	7	18
170	2	19
180	0	15
190	0	11
200	0	4
210	0	1

Answers

1. Load the TWO-WAY template and execute the following:

   ```
   <F>ile<L>oad<L>oadXEx10_2
   <G>raph<E>dit
   2.1<return>                 Enter Low Limit
   2.4<return>                 Enter Hi Limit
   <C>reate<G>auss
   ```

   The graph displayed shows the intersection of the two distributions, both being classic bell-shaped curves. It is obvious that control was tightened. The second curve is taller and narrower. The points distributed near the outside edges are redistributed to the top and center of the curve.

2. Using either a data file or manual entry, enter the data from Exercise 10.3. Once the data are entered, execute the following:

   ```
   <T>est
   <R>egression Quit
   <G>raph
   <R>egression Quit
   ```

   Figure 11.7 shows the created graph.

11 Two-way Graphic Methods

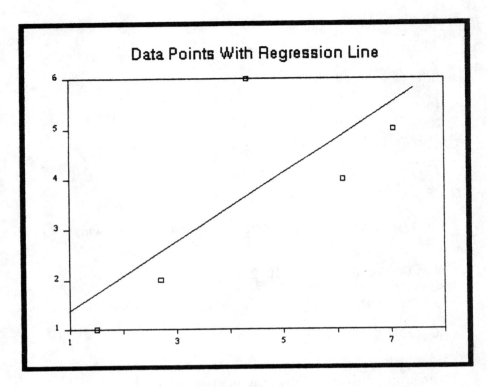

Figure 11.7 Answer to Exercise 11.2

3. Unload the current file from the TWO-WAY template and enter the following:

```
<F>ile<L>oad<L>oadXEx11_1
<G>raph<E>dit
90<return>                    Low Limit
215<return>                   Hi Limit
<C>reate<G>auss<Q>uit
```

One point of intersection of these curves occurs near the center of the graph, at about 150 on the X scale. The second point occurs far off the screen to the left. The Sample Statistics section of the output chart confirms that, in fact, the intersections are at 153 and 48.

Two-way Graphic Methods 11

The Standardized Normal Random Variate (NRV) section of the output expresses these curves in terms of the standard deviations. X1 and X2 relate to the left curve, whose intersections are at -6.51 and +0.98 standard deviations from the mean. The second curve intersects at -6.47 and -0.67 (Y2) standard deviations from the mean.

The probability that a sample from the second distribution will be better than one from the first is the area under the second curve and to the right of the first curve. This area is as follows:

A = 0.5 + .249 - (0.5 - .334) = .583

In other words, when a sample is drawn, the probability is 58.3% that it is better than it would have been if the old process were used. Figure 11.8 shows the curve as it appeared.

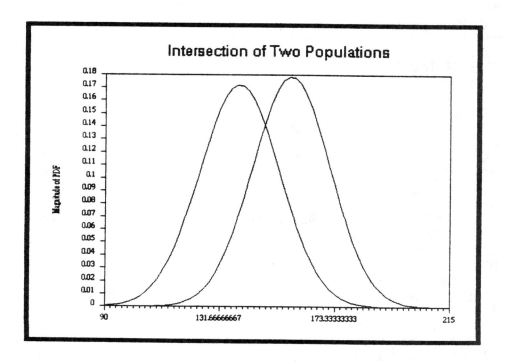

Figure 11.8 Answer to Exercise 11.4

11 Two-way Graphic Methods

TEMPLATE LISTING

Following is a complete listing for the template discussed in this chapter.

```
A13: "Y=
A14: ' Standard err of estimate. .
A15: 'Coefficient of correlation.
A16: ' % of Y variation due to X .
A20: '      Test Level . .
A21: '      Crit Value . .
A22: '      Deg Freedom. .
A23: '      Tails. . . . .
A25: '      Test Value . .
A26: '      Hypothesis . .
A27: '      (Acc=0,Rej=1)
A29:-H29: '*********
B10: '     Points of Intersection
B12: '     Regression Information
B3: "File:
B8: '      Sample X . . . . .
B9: '      Sample Y . . . . .
C13: ^+
C18: "T Test
C19: 'Sml Smpl
C1: '     Two Way Analysis Template
C20: 0.95
C22: @COUNT(C45..C82)+@COUNT(D45..D82)-2
C23: 2
C5: '           Sample Statistics
D18: "Z Test
D19: 'Lrg Smpl
D20: 0.95
D21: @IF(D20=0.95,1.96,@IF(D20=0.975,2.24,@IF(D20=0.99,2.58,@IF
     (D20=0.999,3.29,0))))
D22: (C22)
D23: 2
D7: "Mean
E10: (AB3+AB4)
E13: 'X
E20: '      Test Level . .
E21: '      V1 Value . . .
E22: '      V2 Value . . .
```

continued...

Two-way Graphic Methods 11

...from previous page

```
E23: '    Crit Value . .
E25: '    Test Value . .
E26: '    Hypothesis . .
E27: '      (Acc=0,Rej=1)
E3: 'Operator:
E6: 'Standard Deviation
E7: 'N Weight
F10: (AB3-AB4)
F12: '   Standardized NRV
F13: "X1=
F14: "X2=
F15: "Y1=
F16: "Y2=
F7: 'N-1 Wght
G13: (F2) (E10-D8)/F8
G14: (F2) (F10-D8)/F8
G15: (F2) (E10-D9)/F9
G16: (F2) (F10-D9)/F9
G18: "F Test
G19: "ANOVA
G20: 0.05
G21: +G32-1
G22: +H32-1
G35: @MIN(C37.C38)
G36: @MAX(D37.D38)
G37: (G36-G35)/60
G7: "Count
```

Summary Table Labels

```
A30: 'Label:
A32: 'File:
A35: 'Variable:
A40: 'Graph:
A45: 'Data:
B31: 'Dmnsion
B32: 'Name
B33: 'Descrip
B34: 'Operator
B35: 'Name
```

continued...

333

11 Two-way Graphic Methods

...from previous page

```
B36:  'Units
B37:  'Lo Limit
B38:  'Hi Limit
B39:  'Count
B40:  'Title-1
B41:  'Title-2
B42:  'X-title
B43:  'Y-title
B44:  'Flt Lbl
B45:  'Start
C30:  'Data:
F31:  'Mean
F32:  'Count
F33:  'Std Dev
F34:  'Variance
F35:  'Min
F36:  'Max
F37:  'Cell
F38:  'Roots
G30:  ^X
G35:  @MIN(C38..D38)
G36:  @MAX(C37..D37)
G37:  (G36-G35)/100
G38:  (AB3+AB4)
H30:  ^Y
H38:  (AB3-AB4)
```

Data Range Labels

continued...

Two-way Graphic Methods 11

...from previous page

```
J3:  '*************************************************************
J4:  '*************************************************************
J5:  '***                                                       ***
J6:  '***                                                       ***
J7:  '***        STATISTICAL MODELING WITH LOTUS 1-2-3          ***
J8:  '***                COMPANION DISKETTE                     ***
J9:  '***                                                       ***
J10: '***                                                       ***
J11: '***         A STATISTICAL SOFTWARE PACKAGE                ***
J12: '***                                                       ***
J13: '***                                                       ***
J14: '***                                                       ***
J15: '***                                                       ***
J16: '***                                                       ***
J17: '***                                                       ***
J18: '*************************************************************
J19: '*************************************************************

                       Marquis



I32:  "        Two Way Analysis Techniques
J100: '         STATISTICAL MODELLING WITH LOTUS 1-2-3
K107: 'Enter required adjustments to graph
K108: '        Enter (S)ave
K109: '        Enter Graph name
K110: '        Quit Graph mode (q)
K111: '        Press <Alt/G> to Restart

             Print Labels and Graph Instructions




I34: 'Menu1
I35: 'Test
I36: 'Regression  T-Test  Z-Test  F-Test  Quit
I37: '/xmmenu10~
```

continued...

11 Two-way Graphic Methods

...from previous page

```
J35:  'File
J36:  'Load  Unload  Save  Replace  Directory  Quit
J37:  '/xmmenu2~
K35:  'Print
K36:  'Summary  Test-Results  FormFeed  Quit
K37:  '/xmmenu3~
L35:  'Graph
L36:  'Xscatter  Yscatter  Gauss  Regression  Line  Smooth  Quit
L37:  '/xmmenu4~
M35:  'Cursor
M36:  'Left  Right  Up  Down  GoTo  Quit
M37:  '/xmmenu5~
N35:  'Adjust
N36:  'Sort  Constant  Halt  Quit
N37:  '/xmmenu6~
O35:  'Quit
O36:  'Retrieve Menu Template
O37:  '/fdb:\~
O38:  '/frmenu~
I40:  'Menu2 - File
I41:  'Load
I42:  'Load  LoadY  Table  Quit
I43:  '/xmmenu8~
J41:  'Unload
J42:  'Remove a file from the template
J43:  '/xcunload~
J44:  '/xmmenu1~
K41:  'Save
K42:  'Store the current data file
K43:  '{goto}c31~
K44:  '/fxv{?}~.{pgdn}{end}{down}{right}~
K45:  '{home}/xmmenu1~
L41:  'Replace
L42:  'Save an update to an existing file
L43:  '{goto}c31~
L44:  '/fxv{?}~.{pgdn}{end}{down}{right}~r
L45:  '{home}/xmmenu1~
M41:  'Directory
M42:  'Change the directory
M43:  '/fd{?}~
```

continued...

Two-way Graphic Methods 11

...from previous page

```
M44:  '/xmmenu1~
N41:  'Quit
N42:  'Return to the Primary Menu
N43:  '/xmmenu1~


I46:  'Menu3 - Print
I47:  'Summary
I48:  'Print the summary report
I49:  '/xcsummary~
I50:  '/xmmenu3~
J47:  'Test-Results
J48:  'Print the test results
J49:  '/xctests~
J50:  '/xmmenu3~
K47:  'FormFeed
K48:  'Advance page to top of form
K49:  '/pppq
K50:  '/xmmenu3~
L47:  'Quit
L48:  'Return to main menu
L49:  '/xmmenu1~


I53:  'Menu4 - Graph
I54:  'Create
I55:  'Histogram  Data  Ogive  1-Gauss  2-Gauss  Quit
I56:  '/xmmenu7~
J54:  'Save
J55:  'Save the current graph to disk
J56:  '/xgsave~
K54:  'Edit
K55:  'Edit Graph Limits
K56:  '/xgedit~
L54:  'Quit
L55:  'Return to main menu
L56:  '/ree45.g2048~
L57:  '/xmmenu1~
```

continued...

11 Two-way Graphic Methods

...from previous page

```
I59: 'Menu5 - Cursor
I60: 'Left
I61: 'Move cursor left five columns
I62: '{left}{left}{left}{left}{left}
I63: '/xmmenu5~
J60: 'Right
J61: 'Move cursor right five columns
J62: '{right}{right}{right}{right}{right}
J63: '/xmmenu5~
K60: 'Up
K61: 'Move cursor up one page
K62: '{up}{up}{up}{up}{up}{up}{up}{up}{up}
K63: '/xmmenu5~
L60: 'Down
L61: 'Move cursor down one page
L62: '{down}{down}{down}{down}{down}{down}{down}{down}{down}
L63: '/xmmenu5~
M60: 'GoTo
M61: 'Home   Address
M62: '/xmmenu9~
N60: 'Quit
N61: 'return to main menu
N62: '{home}
N63: '/xmmenu1~


I66: 'Menu6 - Adjust
I67: 'Sort
I68: 'Xup  DownX  UpY  Ydown  Index  Quit
I69: '/xmmenu11~
J67: 'Index
J68: 'Add index column to data
J69: '/xcindex~
J70: '/xmmenu6~
K67: 'Constant
K68: 'Low-Bin  Bin-Size  Chi-Level
K69: '{goto}q12~
K70: '/xmmenu12~
L67: 'Halt
```

continued...

Two-way Graphic Methods 11

...from previous page

```
L68:  'Halt the MACRO to make changes
L69:  '/wtc{home}
L70:  '/xq
M67:  'Quit
M68:  'Return to primary menu
M69:  '/xmmenu1~


Q34:  'Menu7 - Create Graphs
Q35:  'Xscatter
Q36:  'Create a Histogram Graph
Q37:  '/xcxscat~
Q38:  '/xmmenu4~
R35:  'Yscatter
R36:  'Plot Data Points Sequentially
R37:  '/xcyscat~
R38:  '/xmmenu4~
S35:  'Gauss
S36:  'Generate Ogive From Frequency Distribution
S37:  '/xcgauss~
S38:  '/xmmenu4~
T35:  'Regression
T36:  'Generate Gaussian Bar Graph
T37:  '/xcreggraph~
T38:  '/xmmenu4~
U35:  'Smooth
U36:  'Graph with three point smoothing
U37:  '/xcsmooth~
U38:  '/xmmenu4~
V35:  'Quit
V36:  'Return to Primary Menu
V37:  '/xmmenu4~


Q40:  'Menu8 - Load options
Q41:  'Load
Q42:  'Load X column or X,Y columns
Q43:  '/xcload~
Q44:  '/xmmenu1~
```

continued...

11 Two-way Graphic Methods

...from previous page

```
R41:  'LoadY
R42:  'Load Y column only
R43:  '/xcloady~
R44:  '/xmmenu1~
S41:  'Table
S42:  'Load Data into Calculation Table
S43:  '/xctable~
S44:  '/xmmenu1~
T41:  'Quit
T42:  'Return to Main Menu
T43:  '/xmmenu1~


Q46:  'Menu9 - GoTo Options for Cursor
Q47:  'Home
Q48:  'Home Cursor
Q49:  '{home}
Q50:  '/xmmenu1~
R47:  'Address
R48:  'GoTo a specific cell
R49:  '{goto}{?}~
R50:  '/xmmenu9~
S47:  'Quit
S48:  'Return to main menu
S49:  '/xmmenu1~


Q53:  'Menu10 - Test Options
Q54:  'Regression
Q55:  'Compute Regression between X and Y
Q56:  '/xcreg~
Q57:  '/xmmenu10~
R54:  'T-Test
R55:  'Perform Small Sample test on means of X and Y
R56:  '/xcttest~
R57:  '/xmmenu10~
S54:  'F-Test
S55:  'Perform F test on ratio of variances
S56:  '/xcftest~
S57:  '/xmmenu10~
```

continued...

Two-way Graphic Methods 11

...from previous page

```
T54:  'Z-Test
T55:  'Perform Large Sample test on means of X and Y
T56:  '/xcztest~
T57:  '/xmmenu10~
U54:  'Project
U55:  'Project value of Y based on regression
U56:  '/xcfrcst~
U57:  '/xmmenu10~
V54:  'Quit
V55:  'Return to main menu
V56:  '/xmmenu1~


Q59:  'Menu11 - Sort options
Q60:  'Xup
Q61:  'Sort based on X column in ascending order
Q62:  '{goto}c45~/dsrd.{end}{down}{right}{right}~
Q63:  'p.{end}{down}~a~g
Q64:  '/xmmenu11~
R60:  'DownX
R61:  'Sort based on X column in descending order
R62:  '{goto}c45~/dsrd.{end}{down}{right}{right}~
R63:  'p.{end}{down}~d~g
R64:  '/xmmenu11~
S60:  'UpY
S61:  'Sort based on Y column in ascending order
S62:  '{goto}c45~/dsrd.{end}{down}{right}{right}~
S63:  'p.{right}{end}{down}~a~g
S64:  '/xmmenu11~
T60:  'Ydown
T61:  'Sort based on Y column in descending order
T62:  '{goto}c45~/dsrd.{end}{down}{right}{right}~
T63:  'p.{right}{end}{down}~d~g
T64:  '/xmmenu11~
U60:  'Index
U61:  'Resort data to index - original order
U62:  '{goto}c45~/dsrd.{end}{down}{right}{right}~
U63:  'p{right}{right}.{end}{down}~a~g
U64:  '/xmmenu11~
```

continued...

11 Two-way Graphic Methods

...from previous page

```
V60:  'Quit
V61:  'Return to main menu
V62:  '/xmmenu1~


Q66:  'Menu12 - Change Constants
Q67:  'T Test level
Q68:  'Change the T test level
Q69:  '/xnEnter new T test level ~c20~
Q70:  '/xmmenu12~
R67:  'Z Test level
R68:  'Change the Z Test level
R69:  '/xnEnter new Z test level ~d20~
R70:  '/xmmenu12~
S67:  'F Test level
S68:  'Change the F test level
S69:  '/xnEnter new F test level ~g20~
S70:  '/xmmenu12~
T67:  'Quit
T68:  'Return to main menu
T69:  '{home}
T70:  '/xmmenu6~


I23:  'start
I24:  '{home}{goto}i1~{goto}k15~/cI32~~
I25:  '/fdb:\~
I26:  '/xl press Return to continue~K23~{home}
I27:  '/xmmenu1~


I73:  'Load MACRO
I74:  '{goto}c31~
I75:  '/fcce{?}~
I76:  '{goto}c45~
I77:  '/rnddata~/rnddata1~
I78:  '/rncdata~.{end}{down}~
I79:  '{goto}g31~
I80:  '@avg(data)~{edit}{calc}~
I81:  '{down}@count(data)~{edit}{calc}~
```

continued...

Two-way Graphic Methods 11

...from previous page

```
I82: '{down}@std(data)~{edit}{calc}~
I83: '{down}@var(data)~{edit}{calc}~
I84: '/xic31<<>>2~/xg:92~
I85: '{goto}d45~
I86: '/rncdata1~.{end}{down}~
I87: '{goto}h31~
I88: '@avg(data1)~{edit}{calc}~
I89: '{down}@count(data1)~{edit}{calc}~
I90: '{down}@std(data1)~{edit}{calc}~
I91: '{down}@var(data1)~{edit}{calc}~
I92: '/c$g$31~d8~/c$h$31~d9~
I93: '/c$g$33~e8~/c$h$33~e9~
I94: '{goto}f8~($e$8*@sqrt($g$32/($g$32-1)))~
I95: '{down}($e$9*@sqrt($h$32/($h$32-1)))~
I96: '/cg32~g8~/ch32~g9~
I97: '{home}
I98: '/cc32~c3~/cc34~f3~
I99: '/xr
K73: 'LoadY MACRO
K74: '{goto}d31~
K75: '/fcce{?}~
K76: '/xgi85~
N73: 'Unload MACRO
N74: '/reg31.h34~
N75: '/rec31.d2048~
N76: '/rec25.d26~/ree40.h2048~
N77: '/reg23.g26~
N78: '/reb13~/rec3~
N79: '/red13~/ref3~
N80: '/ree14.e16~
N81: '/red8.g9~/rec21~
N82: '/xr
N84: 'Table Subroutine
N85: '/xnEnter Avg1 ~g31~
N86: '/xnEnter Avg2 ~h31~
N87: '/xnEnter Count1 ~g32~
N88: '/xnEnter Count2 ~h32~
N89: '/xnEnted Std Dev1 ~g33~
N90: '/xnEnter Std Dev2 ~h33~
N91: '{goto}g34~({up}^2)~
```

continued...

11 Two-way Graphic Methods

...from previous page

```
N92:  '/c~{right}~
N93:  '/xgi92~


I101: 'Print Summary Subroutine
I102: '/ppcara29.h29~g
I103: 'ri100.p100~g
I104: 'ra1.h16~gra29.h29~gq
I105: '/xr
M101: 'Print test results
M102: '/ppcara18.h23~g
M103: 'ra25.h27~gra29.h29~gq
M104: '/xr


Q2:   'Regression Subroutine
Q4:   '{home}{goto}e45~
Q5:   '({left}{left}*{left})~
Q6:   '{right}({left}{left}{left}^2)
Q7:   '{left}/c.{right}~.{left}{end}{down}{right}~
Q8:   '{up}@sum({down}.{end}{down})~
Q9:   '/c~{right}~
Q10:  '{up}({down}-@count(data)*@avg(data)*@avg(data1))~
Q11:  '{right}({down}-@count(data)*(@avg(data)^2))~
Q12:  '{goto}d13~
Q13:  '(e43/f43)~{edit}{calc}~
Q14:  '{left}{left}
Q15:  '@avg(data1)-@avg(data)*(d13)~{edit}{calc}~
Q16:  '{goto}e43~/re.{right}{end}{down}~
Q17:  '{down}{down}({left}{left}*$d$13+$b$13)
Q18:  '{right}({left}{left}-{left})^2~
Q19:  '{right}(({left}{left}-$h$31)^2)
Q20:  '{right}(({left}{left}{left}{left}-$h$31)^2)~
Q21:  '{left}{left}{left}
Q22:  '/c.{right}{right}{right}~.{left}{end}{down}{right}~
Q23:  '{up}(@sum({down}.{end}{down}))~
Q24:  '/c~.{right}{right}{right}~
Q25:  '{home}{goto}e14~
Q26:  '@sqrt(f44/(g32-2))~{edit}{calc}~
Q27:  '{down}@sqrt(g44/h44)~{edit}{calc}~
```

continued...

Two-way Graphic Methods 11

...from previous page

```
Q28:  '{down}(g44/h44)*100~/rff3~~{edit}{calc}~
Q29:  '{goto}e44~
Q30:  '/re.{end}{down}{right}{right}{right}~
Q31:  '{home}~
Q32:  '/xr


Q73:  'Z Test Subroutine
Q75:  '{home}{pgdn}{goto}a10~{goto}d25~
Q76:  '(g31-h31)/@sqrt((g34/g32)+(h34/h32)){edit}{calc}~
Q77:  '{down}@if(@abs(d25)d21,1,0)
Q78:  '{home}
Q79:  '/xr


Q82:  'T Test Subroutine
Q84:  '{pgdn}{goto}a10~/xnEnter T critical val ~c21~
Q85:  '{goto}e45~
Q86:  '({left}{left}-$g$31)^2~
Q87:  '{right}({left}{left}-$h$31)^2~
Q88:  '{left}/c.{right}~.{left}{end}{down}{right}~
Q89:  '{up}@sum({down}.{end}{down})~
Q90:  '/c~{right}~
Q91:  '{up}@sqrt(({down}+{down}{right})/(G32+H32-2))~
Q92:  '{right}({left}*@sqrt((1/g32)+(1/h32)))~
Q93:  '{goto}c25~(g31-h31)/f43~{edit}{calc}~
Q94:  '{down}@if(@abs(c25)c21,1,0)~
Q95:  '{goto}e43~/re.{end}{down}{right}~{goto}a10~
Q96:  '/xr


Q99:  'F Test Subroutine
Q101: '{home}{goto}a10~
Q102: '/xnEnter F critical value~g23~
Q103: '{goto}g25~(g34/h34)~
Q104: '{down}@if(g25g23,1,0)~
Q105: '{home}
Q106: '/xr
```

continued...

11 Two-way Graphic Methods

...from previous page

```
AA3:  "S1=
AA4:  "S2=
AB3:  -Z4/(2*Z3)
AB4:  (@SQRT((Z4^2)-4*Z3*Z5))/(2*Z3)
Y1:   'Scratchpad
Y3:   "a=
Y4:   "b=
Y5:   "c=
Z3:   ((F8^2)-(F9^2))
Z4:   2*(G31*(F9^2)-H31*(F8^2))
Z5:   (F8*H31)^2-(F9*G31)^2-2*(F8*F9)^2*@LN(F8/F9)


Y9:   'Index Subroutine
Y11:  '{goto}e45~/dfe45~~~~
Y12:  '/df.{left}{left}{end}{down}{right}{right}~1~1~~
Y13:  '/c.{end}{down}~{right}~
Y14:  '/xr~



Y17:  'Smooth Subroutine
Y19:  '/ref45.f2048~{goto}f46~
Y20:  '@sum(d45.d47)/3~
Y21:  '/c~.{left}{left}{end}{down}{right}{right}~
Y22:  '/grgxdata~x{up}{up}~adata1~a{up}{up}~
Y23:  'b.{end}{down}{up}{up}~oss4~
Y24:  'tfRaw Data vs Three Point Smoothed~
Y25:  'ts\c40~tx\c42~ty\c43~qvq
Y26:  '/ree45.h2048~{home}
Y27:  '/ree45.h2048~/xr~



Y29:  'Save - saves graph
Y30:  '{home}{goto}a10~
Y31:  '/wwh{window}{goto}I106~
Y32:  '/g
Y34:  'Restart - using \g
Y35:  '/wwc{home}
```

continued...

Two-way Graphic Methods 11

...from previous page

```
Y36:  '/xmmenu4~


AC39: 'Yscat
AC40: '/grgxdata1~adata~
AC41: 'oss4~
AC42: 'tfY-X Scatter Plot~
AC43: 'ts\c40~tx\d44~ty\c44~qvq
AC44: '/ree45.h2048~{home}
AC45: '/xr~
Y39:  'Xscat
Y40:  '/grgxdata~adata1~
Y41:  'oss4~fasq
Y42:  'tfX-Y Scatter Plot~
Y43:  'ts\c40~tx\c44~ty\d44~qvq
Y44:  '/ree45.h2048~{home}
Y45:  '/xr~


Y47:  'Gauss
Y48:  '{goto}e45~
Y49:  '/dfe45.e105~g35~g37~g36~
Y50:  (F2)'{right}(@EXP((-({left}-
      $d$8)^2)/(2*($f$8^2))))/$f$8*@SQRT(2*@PI)~
Y51:  (F2)'{right}(@EXP((-({left}{left}-
      $d$9)^2)/(2*($f$9^2))))/$f$9 *@SQRT(2*@PI)~
Y52:  '/cf45.g45~f45.f105~
Y53:  '/grgx{left}{left}.{end}{down}~a{left}.{end}{down}~b.{end}{down}
Y54:  'otfIntersection of Two Populations~ts\c40~
Y55:  'tx\c35~tyMagnitude of PDF~q
Y56:  'oss20~falblqqvq
Y57:  '/ree45.h2048~{home}
Y58:  '/xr~


Y61:  'Regression graph
Y62:  '{goto}e45~
Y63:  '({left}{left}*$d$13+$b$13)~
Y64:  '/c~.{left}{end}{down}{right}~
```

continued...

11 Two-way Graphic Methods

...from previous page

```
Y65:  '/grgx{left}{left}.{end}{down}~
Y66:  'a{left}.{end}{down}~
Y67:  'b.{end}{down}~
Y68:  'ofasblqss5~
Y69:  'tfData Points With Regression Line~
Y70:  'ts\c40~tx\c42~ty\c43~qvq
Y71:  '/ree45.h2048~{home}
Y72:  '/xr~
Y75:  'frcst
Y76:  '{goto}y100~
Y77:  /xn enter x value~ac100~
Y78:  '{goto}ac101~(b13 + (d13 * ac100))~
Y79:  'xl press Return to continue ~ac100~
Y80:  '/reac100.ac101~
Y81:  '{home}~/xr~

AA100: 'for an X of
AA101: 'the projected Y is

AC61: 'edit
AC62: '/xn enter now low limit~c37~
AC63: '/xn enter new hi limit~c38~
AC64: '/cc37.c38~d37~
AC65: '/xmmenu4~
```

CHAPTER 12

NONPARAMETRIC METHODS

12 Nonparametric Methods

All methods shown so far are valid for samples from one of several well-characterized distributions. In fact, most statistical work is done on samples representative of either a normal, chi-square, student, or binomial distribution. Other models are often used, however, and random distributions do exist. Frequently, nonrandom patterns occur that do not meet the criterion for analysis using the standard models. In these cases, **nonparametric** — or distribution-independent — methods are used.

Three widely applicable methods are presented in this chapter: the **runs** test, the **Mann-Whitney** test, and the **sign** test. The **runs** test provides a measure of a data set's randomness. In a controlled statistical experiment, special methods are employed to ensure that the data generated are actually random; these methods can be costly and time-consuming. The runs test provides a method for assessing the randomness of any data set. If the result is negative, the data can be used as if they were gathered statistically. The **Mann-Whitney** test — like the Z and T tests presented earlier — provides a measure of significance for the assumption that two samples come from the same universe. Unlike the other tests, the Mann-Whitney test (often called the U test) does not make the assumption that the universe or sample populations are normally distributed. The **sign** test also measures the hypothesis that two sample populations are from the same universe. The sign test — like the T test — is suited for smaller samples. It also requires pairs of observations.

Figure 12.1 is a sample of the output from this template. Under the heading are three sections providing the results of the runs, sign, and Mann-Whitney tests, respectively.

```
              STATISTICAL MODELLING WITH LOTUS 1-2-3
                     Non-parametric Test Methods

    File:                    Operator:

    Runs Test            Sign Test              Mann-Whitney

    Level     0.05       Level     0.05         Level     0.05
    N1                   Px
    N2                                          Ratio     1.96
    Runs                 Ratio     1.64         N1
                         N                      N2
    Ratio     1.96       Py                     Rnk Sum
    Er                   z                      U
    Sr                   Hypoth                 Su
    z                    (Acc=0,Rej=1)          z
    Hypoth                                      Hypoth
    (Rnd=0)                                     (Acc=0,Rej=1)
```

Figure 12.1 Sample Output

Nonparametric Methods 12

Figure 12.2 shows the layout for the Nonparametric Methods template. This is the simplest template in the book, with only three screens and five menus.

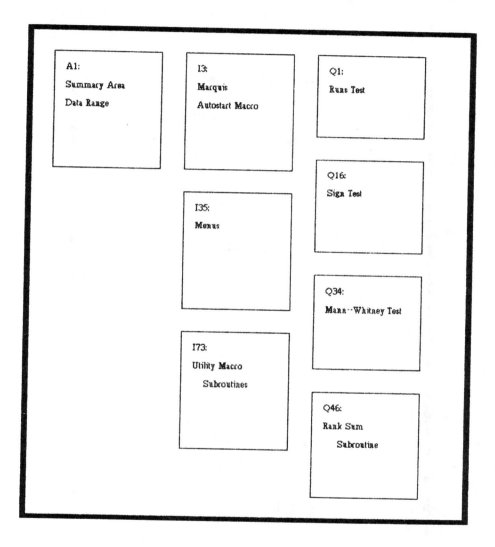

Figure 12.2 Nonparametric Template Block Diagram

12 Nonparametric Methods

MENUS

Since there are only three methods presented in this chapter, the menu is quite simple. Figure 12.3 shows the menu in tree structure. There are a few minor changes in the way this template operates in comparison to earlier templates.

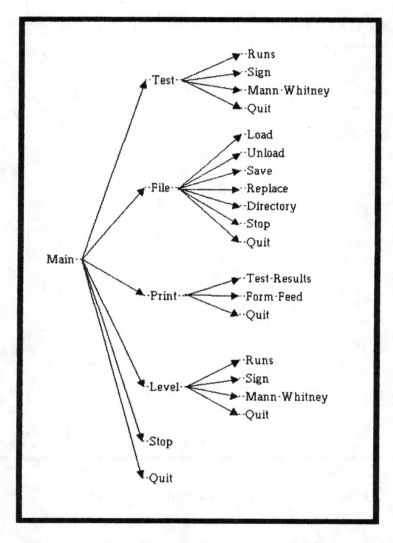

Figure 12.3 Menu Structure for Nonparametric Template

Nonparametric Methods **12**

Main Menu

The Stop function has been moved to the first menu; it halts execution of the macro and allows the analyst to manually operate on or move around the spreadsheet. There is no need for cursor mobility commands because there are no permanent tables or summary zones not shown on the screen. As with previous templates, an <Alt/A> command restarts the macro operation.

File Menu

File save and **replace** operations are provided to allow updating of the disk files after modification of data on the screen. Raw data can be entered directly into the template, bypassing the file load operation completely.

Test Menu

The Test menu allows the user to select and run one or more of the tests. Some user data input is required in executing these tests. Options include **Sign**, **Runs**, **U-test**, and **Quit** to return to the main menu.

Print Menu

The Print menu has only one real function — to output the screen contents to the printer. In most cases, the data will be incorporated into another chart or report. This whole section could easily have been eliminated in favor of the <shift><prtsc> function built into the computer, but it was included for completeness.

METHODS AND APPLICATIONS

Runs Test

The **runs** test can be applied to either numerical data or attributes data as a measure of randomness. A series of experiments is run, and the results are recorded as successes or failures. Here, success and failure are the two arbitrary possible outcomes. Numerically, these outcomes are frequently taken as values above and below the median for small samples. The runs test determines whether the variation in results is random or the result of a bias in the data.

12 Nonparametric Methods

A subtle advantage of the runs test is its ability to test data sets for randomness to determine if they make a relevant sample. Generating valid random samples can be very costly. If an empirical data set is found, it can be tested for randomness. If the runs test index is low, the data set can be used as if it were selected using a statistically random method.

In the following example, a manufacturer wants to test his product head-to-head with his competitor's. Fifty people are given samples and allowed to choose between the two products. Their responses are tabulated below with either a "c" or a "P" representing the choice. The result is to be accepted only if it is found that the data are random. The hypothesis being tested is as follows:

H_O: The sample evaluated is random in nature.

Data are recorded in the order taken and clustered to allow the groups or **runs** to be counted:

cccc PPP c PP ccc PP cc PP cc P c P ccc
PPP ccccc PP cccc PPP c PPP cc

Three pieces of information are extracted from this list: the number of c's, the number of P's, and the number of runs.

$Nc = 28$
$Np = 22$
$R = 21$

From these amounts, three more calculations can be made. Er is the expected number of runs, based on the numbers of successes and failures:

$Er = (2N_1N_2/(N_1 + N_2)) + 1$

Equation 12.1

Nonparametric Methods 12

The standard deviation of the number of runs — S_r — is also a function of the two quantities:

$$S_r = (2N_1N_2(2N_1N_2 - N_1 - N_2)/(N_1+N_2)^2(N_1 + N_2 - 1))^{0.5}$$

Equation 12.2

Finally, the inevitable test statistic to be compared to the standard normal random variate (NRV) is as follows:

$$z = (R-E_r)/S_r$$

Equation 12.3

Substitute the values from the example into these three equations:

$$E_r = ((2*28*22)/(28*22)) + 1 = 25.64$$

$$S_r = (2*28*22(2*28*22 - 28 - 22)/(28 + 22)^2(28 + 22 - 1)^{0.5} = 3.4478$$

$$z = (R - E_r)/S_r = 1.345$$

The value for z obtained previously is compared to the NRV from the table of areas under the normal curve. For a test level of 0.05 on a two-tailed test, the coefficient for comparison is 1.96. (Find 1.96 in the table. The area associated is .475 under each curve [excluding the tail], or .95 overall.) The hypothesis that the distribution is random is accepted because

$$z = 1.3456 < 1.96.$$

12 Nonparametric Methods

Sign Test

The **sign** test operates on pairs of data points taken from populations of non-Gaussian or unknown distribution. With a large sample and normal population, this test approaches the paired T test in accuracy. For less than ten pairs of points, the accuracy is acceptable. Each pair of points should be two treatments of the same variable.

For each pair of points from the two groups (A and B), it is necessary to record the number of times A exceeds B and vice-versa. Ties are not counted in either group, so the number of points recorded in the sign test output may be smaller than the runs or U test outputs. It is necessary to estimate the probability that A will exceed B to find the standard deviation. The following example will show how this is accomplished.

Case Study Five: Manufacturing

A manufacturer of Widgets has a new assembly procedure that will increase the number of units an assembler can build in a week. To test the process, 12 assemblers use the new procedure for one week. Table 12.1 gives the output for these 12 employees, using the old method (column A) and the new method (column B) for one week each. The value in column C is 1 if A > B, and the value in column D is 1 if B > A. Only one column can be marked; for ties, neither is marked. The hypothesis to be tested is as follows:

H_0: The new procedure improved operator outputs.

Performing this calculation requires two treatments. In this case, each line represents a single operator. Each treatment is a use of one of the two methods. If 24 operators had been used, each line would represent two variables with one treatment each—namely, operator and method. In this case, the results would be unacceptable.

Nonparametric Methods 12

A	B	C	D
47	42	1	
38	41		1
35	35		
46	41		1
52	55		1
28	37		1
61	62		1
56	54	1	
48	44	1	
59	52	1	
73	71	1	
58	54	1	
		7	4

Table 12.1

From this data, the following can be extracted:

$Na = 7$ Frequency $A > B$
$Nb = 4$ Frequency $B > A$
$N = Na + Nb = 11$
$X = Max(Na, Nb) = 7$ Larger value of Na and Nb
$p = q = .5$ Probability that $A > B$ (assumption)

These values provide the data for the two equations in this section: the standard deviation,

$$s = (npq)^{0.5}$$

Equation 12.4

and the test statistic,

$$z = (X - np)/s$$

Equation 12.5

357

12 Nonparametric Methods

This is the same form as previously presented. In the numerator, X is the actual value, and np is the expected value. These values are divided by the standard deviation of the estimate.

Substitute the actual values from the example into Equations 12.4 and 12.5:

$$s = (np8)^{0.5} = (11*.5*.5)^{0.5} = 1.658$$

$$z = (X-np)/s = (7-11*.5)/1.658 = 3.920$$

For an alpha value (error) of 5%, the critical value is +/- 1.96. This is the value corresponding to an area of .475 in the table of areas under the standard normal probability density function. In this example, 3.920 is larger than the critical value, so the hypothesis is rejected. The new method did not improve the output.

Mann-Whitney Test

The **Mann-Whitney** test (or U test) is the most accurate of the three tests presented in this chapter because it takes into account magnitudes instead of just signs. The ranks of the data values are established and summed for each data group as part of the calculation, which is why this method is sometimes called the **rank-sum** method. As with the Z test presented earlier, with the Mann-Whitney test it is not necessary to have equal sample sizes in the two groups, but there should be at least ten to twelve samples in each. This test is frequently used with one large group and one small group.

Case Study Six: Softball

Table 12.2 contains data collected by the manager of a softball team. Since this is only an illustration of the Mann-Whitney method, a very small sample is used. Group A consists of batting averages of team members using metal bats. Group B's team members use wood bats. The hypothesis to be evaluated is that there is no difference in batting averages as the result of bat type.

H_0: There is no difference.

Nonparametric Methods 12

A	B
301	180
289	256
340	214
320	244
237	220

Table 12.2

The first step in performing a Mann-Whitney analysis is calculating the rank sum, which requires three operations for a spreadsheet implementation. Figure 12.4 shows the two data sets entered into a spreadsheet. The columns are split so that only one value occurs in a row. The third column is an index added to the starting values. The sole purpose of this column is to allow the data to be returned to their original order after the calculations are done. The final column is the sum of columns A and B. This column is sorted to put the values in rank order.

```
A: 301                                              READY

         A       B       C       D       E       F       G
  1     301              1      301
  2     289              2      289
  3     340              3      340
  4     320              4      320
  5     237              5      237
  6            180       6      180
  7            256       7      256
  8            219       8      213
  9            344       9      344
 10            220      10      220
```

Figure 12.4 Two Data Sets Entered

12 Nonparametric Methods

After the table is sorted in ascending order, based on column D, another index column is added. Figure 12.5 shows the result of that operation. At this point, each element in columns A and B is associated with its rank from lowest to highest. All that remains is to sum the ranks of the elements in column A.

	A	B	C	D	E	F
1		180	6	180	1	
2		213	8	213	2	
3		220	10	220	3	
4	237		5	237	4	
5		256	7	256	5	
6	289		2	289	6	
7	301		1	301	7	
8	320		4	320	8	
9	340		3	340	9	
10		344	9	344	10	

Figure 12.5 Rank Sum Calculation Completed

Summing the ranks can be accomplished easily if the elements are contiguous. Getting the elements to be contiguous is easy if one more sort is performed. The data base is sorted on column A in descending order, which puts all the blank elements at the bottom. The desired rank sum is the sum of column D from the top to the row containing the bottom element in column A. In this example, the top five elements in column D, whose sum is 34, make up the desired rank sum. Figure 12.6 shows the result of this final operation.

```
A1: 340                                                  READY

              A           B           C           D           E           F
    1        340                      3          340          9
    2        320                      4          320          8
    3        301                      1          301          7
    4        289                      2          289          6
    5        237                      5          237          4
    6                    180          6          180          1
    7                    213          8          213          2
    8                    220         10          220          3
    9                    256          7          256          5
   10                    348          9          344         10
```

Figure 12.6 Sorted Table

Now that the table is ordered and complete, the three required values can be extracted:

$N_a = 5$ Number of elements in column A
$N_b = 5$ Number of elements in column B
$R_a = 34$ Sum of the ranks of elements in A

The three equations for the Mann-Whitney test are similar to those of earlier sections. The first is for the U statistic itself:

$$U = N_1 N_2 + N_1(N_2 + 1)/2 - R_1$$

Next, the expected value for the U statistic is as follows:

$$\bar{X} = N_1 N_2 / 2$$

12 Nonparametric Methods

And finally, the standard deviation of U is as follows:

$$S_u = (N_1 N_2 (N_1 + N_2 + 1) / 12)^{0.5}$$

In a manner analogous to previous hypothesis tests, the z statistic is calculated to determine how many standard deviations separate the U statistic and its expected value.

$$z = (U - \overline{X}) / S_u$$

In this example, the absolute value of z is 1.3578, which is compared to the standard value of 1.96 for a two-tailed analysis at the 0.05 level. The hypothesis is accepted. Statistically, there is no difference between bat types.

The three macros are fully described next, along with the required subroutines. The equations for the sample problem are as follows:

$$U = 5*5 + 5(5 + 1)/2 - 34 = 6$$

$$\overline{X} = 5*5/2 = 12.5$$

$$S_u = (5*5(5 + 5 + 1)/12)^{0.5} = 4.787$$

$$z = (6 - 12.5)/4.787 = -1.3578$$

COMMANDS USED

/c	copy range
/df	fill range
/ds	sort range
/fdb:	change to directory B:
/fr	retrieve file
/fxv	file extract (values option)

continued...

Nonparametric Methods 12

...from previous page

/m	move a range
/re	erase a range
/ppca	print printer clear all
/xc	Macro gosub
/xi	Macro If-then
/xm	Macro menu
/xn	Macro – input number from keyboard to cell
/xq	Macro – halt execution
@count (range)	enter the number of elements in range
@if (x,t,v)	enter conditional cell value
@,max (range)	enter highest value in range

MACRO DESCRIPTIONS

Load Subroutine

Only two values are transferred from the data file to the summary screen — the file name and operator (Line I76):

```
I73:  'Load Subroutine
I75:  '{goto}c22~/fcce{?}~
I76:  '/cc23~c4~/cc25~f4~
I77:  '{home}
I78:  '/xr~
```

Print Subroutine

Only one Print subroutine is required. Users may want to break this subroutine into individual routines for each of the three test results.

```
I80:  'Print Subroutine
I82:  '/ppcara20.h20~gll
I83:  'ra1.h20~glllq
I84:  '/xr~
```

12 Nonparametric Methods

Const Subroutine

This subroutine is unique among all others in the book. Be sure to follow the logic carefully. Most programming languages provide a mechanism for passing data from subroutines back and forth by using common memory blocks or by tagging variables to the subroutine call, as has been done here.

The actual subroutine doesn't start until cell I99, where the range name is located. The three cells above it are reserved for passing information. Including these locations in the subroutine area makes them easy to follow, and it allows this subroutine to be called from three different places in the program and return different values each time.

```
I86:  'Const Subroutine
I88:  0.05
I89:  1.96
I90:  2
```

Two values are placed in the reserved area above by keyboard inputs at the time of execution.

```
I91:  '/xnEnter new test level ~i88~
I92:  '/xnEnter number of tails ~i90~
```

There are three possible end locations for this subroutine, based on what kind of calculations must be made. If one of the three standard values is requested, the following three statements jump around the first subroutine ending.

```
I93:  '/xii88=.05~/xgi98~
I94:  '/xii88=.025~/xgi98~
I95:  '/xii88=.01~/xgi98~
```

Nonparametric Methods 12

The first end is a direct entry of a value from the table (provided by the analyst). The value is stored in I89, and the subroutine return is executed.

```
I96:   '/xnEnter z from Normal table ~i89~
I97:   '/xr~
```

If one of the standard values is entered, cell I89 is loaded from either I100 or I103, depending on the value in cell I90, which provides the proper areas based on the number of test tails. Both remaining branches end with a return (/xr). The If statement (/xi) in I99 determines which of the two terminal branches is taken based on the value in I90.

```
I98:    '{goto}i89~
I99:    '/xii90=1~/xgi103~
I100:   '@if(i88=.05,1.96,@if(i88=.025,2.24,@if(i88=.01,2.575,@na)))~
I101:   '{edit}{calc}~{home}/xr~
I103:   '@if(i88=.05,1.64,@if(i88=.025,1.96,@if(i88=.01,2.33,@na)))~
I104:   '{edit}{calc}~{home}/xr~
```

Unload Subroutine

The Unload subroutine erases the data from the table in the lower part of the left columns and the expendable values from the summary table.

```
M73:   'Unload Subroutine
M75:   '/rec22.d2047~
M76:   '/rec4~
M77:   '/ref4~
M78:   '/rec9.c11~
M79:   '/rec14.c17~
M80:   '/ree9~
M81:   '/ree12.e15~
M82:   '/reg11.g17~
M83:   '/xr~
```

12 Nonparametric Methods

Runs Test

The runs test is unique in that none of the data can be entered from a data file. All inputs are from the keyboard through the use of macro input commands. Lines Q3-Q5 perform those inputs.

```
Q1:     'Runs Test
Q3:     '/xnEnter the number of Successes ~c9~
Q4:     '/xnEnter the number of Failures ~c10~
Q5:     '/xnEnter the number of Runs ~c11~
```

Lines Q6-Q8 enter the three equations into the summary report area: Equations 12.1, 12.2, and 12.3.

```
Q6:     '{goto}c14~(1+(2*C9*C10/(C9+C10)))~
Q7:     '{down}(@SQRT((2*C9*C10*(2*C9*C10-C9-C10))/(((C9+C10)^2)*
        (C9+C10-1))))~
Q8     :'{down}((C14-c11)/C15)~
```

The value of z from Equation 12.3 is compared to the standard NRV value in the final line of this macro. If the hypothesis is rejected, a value of one is entered.

```
Q9:     '{goto}c17~@if(@abs(c16)c13,1,0)~
Q10:    '{home}
Q11:    '/xr~
```

Sign Test

The sign test requires one keyboard input — the probability that a positive sign will be encountered. That value is stored in E9, and its compliment is stored in E13.

```
Q14:    'Sign test
Q16:    '{goto}e36~
Q17:    '/xnEnter probability that sign is pos. ~e9~
Q18:    '(1-e9)~{edit}{calc}~/c~e13~/re~
```

Nonparametric Methods 12

The next three lines add two columns in the data area of the spreadsheet. They occur next to the raw data columns. The first column contains a 1 if A>B; otherwise, it contains a zero. The second column is positive where B>A. Both columns are zero for A=B.

```
Q19:   '@if({left}{left}{left},1,0)~
Q20:   '{right}@if({left}{left}{left}{left}{left},1,0)~
Q21:   '{left}/c.{right}~.{left}{end}{down}{right}~
```

The sums of the two signs columns are computed. The total of these two columns is recorded in E12. The {edit}{calc} in Q24 converts this sum into a value so the table can be erased.

```
Q22:   '{up}@sum({down}.{end}{down})~
Q23:   '/c~{right}~
Q24:   '{goto}e12~(e35+f35)~{edit}{calc}~
```

Q26 computes the Z value for the sign test and stores its value in E14. (See Equation 12.5.) Q28 compares the Z value to the standard normal random variate.

```
Q25:   '{down}{down}
Q26:   '((@max(e35.f35)-(e9*e12))/(@sqrt(e9*e12*e13)))~{edit}{calc}~
Q27:   '/rff4~~{goto}e15~
Q28:   '@if(e14e11,1,0)~
Q29:   '{home}/ree35.f2047~
Q30:   '/xr
```

Mann-Whitney Test

The subroutine called in Q34 computes the rank sum for the data set. Details on the computation were provided earlier. The subroutine is explained next.

```
Q32:   'Mann-Whitney Test
Q34:   '/xcrs~
```

12 Nonparametric Methods

The number of elements in the two data ranges is computed and stored in the summary area.

```
Q35:   '{goto}e36~@count({left}{left}{end}{down})~{edit}{calc}~/c~g11~
Q36:   '@count({left}.{end}{down})~{edit}{calc}~/c~g12~/re~
```

Finally, the three equations for the Mann-Whitney test are solved, and the hypothesis comparison is made.

```
Q37:   '{goto}g14~
Q38:   '(g11*g12+(g11*(g11+1)/2)-g13)~
Q39:   '{down}(g11*g12*(g11+g12+1)/12)~
Q40:   '{down}(g14-(g11*g12/2))/@sqrt(g15)~
Q41:   '{goto}g17~@if(@abs(g16)g10,1,0)~{home}
Q42:   '/xr~
```

Rank Sum Subroutine

Lines Q46-Q49 create the spreadsheet described in Figure 12.4. The second data column is moved to one cell below the A column in Q47. Q48 creates the index column, and Q49 makes the column of elements for the rank sort.

```
Q44:   'Rank Sum Subroutine
Q46:   '{goto}d36~
Q47:   '/m{end}{down}~{left}{end}{down}{down}{right}~
Q48:   '{right}/dfe36~1~1~~/df.{left}{end}{down}{end}{down}{right}~~~~
Q49:   '{right}({left}{left}+{left}{left}{left})~/c~.{left}{end}
       {down}{right}~
```

Q50 to Q54 performs the sort and data fill to bring the spreadsheet up to Figure 12.5. The data fill is executed twice, both here and in Q48 above. The first time sets the fill range to the first cell in the desired range. The second fill command can then use the cursor movement commands to find the end of the range. If the command does not start in the proper cell, these cursor movement commands are inoperative.

Nonparametric Methods 12

Q50: '{left}{left}{left}/dsrd.{right}{right}{right}{end}{down}{right}~
Q51: 'p{right}{right}{right}{end}{down}~a~g
Q52: '{right}{right}{right}{right}/dfg36~~~~
Q53: '/df.{left}{end}{down}{right}~~~~
Q54: '{goto}c36~

The final sort accomplishes the state described in Figure 12.6. From here, lines Q57-58 calculate the rank sum. The range for the summation starts at the top of the column. After the anchor (.), the range is extended left to column A, down to the end of that column, and then back to the right to column D. This provides a summation of the elements in D, starting at the top and ending at the row equal to the last value in column A, which is the sum of the ranks of the values in A.

Q55: '/dsrd.{right}{right}{right}{end}{down}{right}~
Q56: 'p.{right}{right}{end}{down}{left}{left}~d~g
Q57: '{goto}g35~@sum({down}.
Q58: '{left}{left}{left}{left}{end}{down}{right}{right}{right}{right})
Q59: '{edit}{calc}~/c~g13~/re~~

The remaining cells erase the scratchpad columns and return the data values to the starting condition.

Q60: '{goto}c36~/dsr
Q61: 'd.{right}{right}{end}{down}~p{right}{right}.{end}{down}~a~g
Q62: '{right}{right}/re.{right}{right}{end}{down}~
Q63: '{left}{end}{down}/m{end}{down}~{end}{up}{down}~
Q64: '{home}/xr~

12 Nonparametric Methods

RANGE NAMES

Using the / <R>ange <N>ame <C>reate function, you must name the following ranges to ensure spreadsheet operation.

Name	Range
Const	I91
Load	I75
Menu1	I36-N37
Menu2	I42-N43
Menu3	I48-K49
Menu4	I54-L55
Menu5	I63-L64
MW	Q34
Print	I82
RS	Q46
Runs	Q3
Sign	Q16
Unload	M75
\0	I24
\A	I24

SUMMARY

- Populations with unknown or unusual distributions can be compared with **nonparametric** tests.

- The **runs test** provides an indication of the randomness of a data set.

- Runs tests can be executed on attributes data.

- The **sign test** provides an indication of the correlation of the pairs in a data set. It will indicate if they come from the same population.

- The **Mann-Whitney test** is based on the rank sum principle. It indicates whether or not the means of two populations are equal, independent of distribution.

- Mann-Whitney is always a two-tailed test.

12 Nonparametric Methods

EXERCISES

Case Study Seven: Education

A school is interested in comparing performance of substitutes with that of regular teachers. Column A in Table 12.3 represents test scores for a group of students after one unit of instruction. Column B represents scores for the class after one unit of instruction by a substitute teacher. The following exercises relate to the data in this table.

A	B
Teacher	Substitute
88	78
85	84
79	83
58	64
61	87
74	79
99	100
94	94
92	96
100	98
97	94
68	69
53	59
61	68
44	67
75	55
66	76
65	59
67	66
63	58
64	59
71	65
74	77
76	78

Table 12.3

12 Nonparametric Methods

1. Based on the number of students who improved, the number of students who did worse, and the pattern, is this a statistically random sample? What was the expected number — and standard deviation — of runs?

2. Perform a sign test at the .05 level; assume the probability of change to be 0.5. Interpret the results.

3. Perform a Mann-Whitney U Test at the 0.05 level. Interpret the results.

Answers

1. The hypothesis to be tested is as follows:

    H_O: There is a nonrandom pattern.

    It is necessary to create a column after table 12.3 that identifies each pair as a success or failure. In this problem, a success will be considered to be a higher score on the second test. This table is shown, in part, below:

    ```
    88    78    -
    85    84    -
    79    83    +
    58    64    +
    . . . .
    ```

    The column created is summarized and partitioned:

    -- + + + + + -- + + - + - + -- + -- + +

372

There are thirteen successes (+), ten failures (-), and one tie that does not appear. There are twelve partitions or groups of data.

Now that the input data are ready, enter 1-2-3 and load the Non-Parm Template. Enter

```
Test Runs
13 <return>        Number of successes
10 <return>        Number of failures
12 <return>        Number of runs (partitions)
```

When the computations have been completed, the screen will return to the main menu, and the results will be displayed. The first column summarizes the results of the runs test. The expected number of runs was 12.3, with a standard deviation of 2.3. The Z value calculated is 0.132266. When Z is compared to the standard test ratio 1.96, the hypothesis is rejected. The pattern is statistically random (since the actual number of runs is close to the expected number). The hypoth value in the display is set to zero.

2. The hypothesis to be tested is as follows:

H_0: The substitute teacher produces higher test results.

Quit test mode to return to the main menu. Enter

```
<F>ile <L>oad Ex12 <return>
<T>est <S>ign
0.5 <return>              Probable equality = 0.5
```

The second column of data contains the results for the sign test. The computed z value of 0.6255 is less than the test ratio 1.64, so the hypothesis is rejected. The substitute teacher did not produce statistically higher test results.

12 Nonparametric Methods

3. The final test is the Mann-Whitney, which tests the same hypothesis as in problem #2. Enter:

    <M>ann-Whitney

The third data column provides the results. The rank sum is 570, the value of the U statistic is 306, and the standard deviation about U is 48.49. The resulting z value is 0.371153, which is significantly less than 1.96, confirming the rejection of the hypothesis. Remember that the U test tends to be more accurate for large samples and is the only method presented for samples of unequal size.

Figure 12.7 shows the screen as it appears after running the three exercises in this chapter.

```
              STATISTICAL MODELLING WITH LOTUS 1-2-3
                      Non-parametric Test Methods
       File:               Operator:

          Runs Test           Sign Test           Mann-Whitney

       Level    0.05       Level    0.05       Level    0.05
       N1         13       Px        0.5
       N2         10                           Ratio    1.96
       Runs       12       Ratio    1.64       N1         24
                           N           23      N2         24
       Ratio    1.96       Py        0.5      Rnk Sum    570
       Er  12.30434         z      0.6255      U          306
       Sr  2.301026        Hypoth      0       Su       48.49
        z  0.132266        (Acc=0,Rej=1)        z    0.371153
       Hypoth      0                           Hypoth      0
       (Rnd=0)                                 (Acc=0,Rej=1)
```

Figure 12.7 Answer to Exercises

Nonparametric Methods 12

TEMPLATE LISTING

Following is the complete listing for the template in this chapter.

```
A20:-H20: '*********
B1:  '        STATISTICAL MODELLING WITH LOTUS 1-2-3
B4:  "File:
B6:  '     Runs Test
B8:  "Level
B9:  "N1
B10: "N2
B11: "Runs
B13: "Ratio
B14: "Er
B15: "Sr
B16: "z
B17: "Hypoth
B18: "     (Rnd=0)
C2:  '     Non-parametric Test Methods
C8:  0.05
C13: 1.96
D6:  '     Sign Test
D8:  "Level
D9:  "Px
D11: "Ratio
D12: "N
D13: "Py
D14: "z
D15: "Hypoth
D16: '    (Acc=0,Rej=1)
E4:  'Operator:
E8:  0.05
```

continued...

12 Nonparametric Methods

...from previous page

E11: 1.64
F6: '   Mann-Whitney
F8: "Level
F10: "Ratio
F11: "N1
F12: "N2
F13: "Rnk Sum
F14: "U
F15: "Su
F16: "z
F17: "Hypoth
F18: '   (Acc=0,Rej=1)
G8: 0.05
G10: 1.96

Summary Range Labels

A21: 'Label:
A23: 'File:
A26: 'Variable:
A31: 'Graph:
A36: 'Data:
B22: 'Dmnsion
B23: 'Name
B24: 'Descrip
B25: 'Operator
B26: 'Name
B27: 'Units
B28: 'Hi Limit
B29: 'Lo Limit

continued...

...from previous page

```
B30: 'Count
B31: 'Title-1
B32: 'Title-2
B33: 'X-title
B34: 'Y-title
B35: 'Flt Lbl
B36: 'Start
C21: 'Data:
```

Data Range Labels

```
I21: 'MACROs
J3:  '****************************************************
J4:  '****************************************************
J5:  '***                                              ***
J6:  '***                                              ***
J7:  '***      STATISTICAL MODELING WITH LOTUS 1-2-3   ***
J8:  '***              COMPANION DISKETTE              ***
J9:  '***                                              ***
J10: '***                                              ***
J11: '***        A STATISTICAL SOFTWARE PACKAGE        ***
J12: '***                                              ***
J13: '***                                              ***
J14: '***                                              ***
J15: '***                                              ***
J16: '***                                              ***
J17: '***                                              ***
J18: '****************************************************
J19: '****************************************************
K15: '     Non-Parametric Methods Template
```

Marquis and other labels

```
I35: 'Menu1
I36: 'File
I37: 'Load   Unload   Save   Replace   Directory   Quit
I38: '/xmmenu2~
J36: 'Test
J37: 'Runs   Sign   Mann-Whitney   Quit
J38: '/xmmenu5~
K36: 'Print
```

continued...

12 Nonparametric Methods

...from previous page

```
K37:  'Test-results  Form-feed  quit
K38:  '/xmmenu3~
L36:  'Level
L37:  'Runs  Sign  Mann-Whitney  Quit
L38:  '/xmmenu4~
M36:  'Stop
M37:  'Halt Macro execution
M38:  '/xq
N36:  'Quit
N37:  'Retrieve Menu Template
N38:  '/fdb:\~
N39:  '/frmenu~


I41:  'Menu2 - File
I42:  'Load
I43:  'Load a file from disk
I44:  '/xcload~
I45:  '/xmmenu1~
J42:  'Unload
J43:  'Remove a file from the template
J44:  '/xcunload~
J45:  '/xmmenu1~
K42:  'Save
K43:  'Store the current data file
K44:  '{goto}c22~
K45:  '/fxv{?}~.{pgdn}{end}{down}{right}~
K46:  '{home}/xmmenu1~
L42:  'Replace
L43:  'Save an update to an existing file
L44:  '{goto}c22~
L45:  '/fxv{?}~.{pgdn}{end}{down}{right}~r
L46:  '{home}/xmmenu1~
M42:  'Directory
M43:  'Change the directory
M44:  '/fd{?}~
M45:  '/xmmenu1~
N42:  'Quit
N43:  'Return to the Primary Menu
N44:  '/xmmenu1~
```

continued...

Nonparametric Methods **12**

...from previous page

```
I47:  'Menu3 - Print
I48:  'Test-Results
I49:  'Print the test results
I50:  '/xcprint~
I51:  '/xmmenu3~
J48:  'FormFeed
J49:  'Advance page to top of form
J50:  '/pppq
J51:  '/xmmenu3~
K48:  'Quit
K49:  'Return to main menu
K50:  '/xmmenu1~


I53:  'Menu4 - Change Constants
I54:  'Runs
I55:  'Change Runs test level
I56:  '/xcconst~
I57:  '/ci88~c8~
I58:  '/ci89~c13~
I59:  '/xmmenu4~
J54:  'Sign
J55:  'Change Sign test level
J56:  '/xcconst~
J57:  '/ci88~e8~
J58:  '/ci89~e11~
J59:  '/xmmenu4~
K54:  'Mann-Whitney
K55:  'Change Mann-Whitney test level
K56:  '/xcconst~
K57:  '/ci88~g8~
K58:  '/ci89~g10~
K59:  '/xmmenu4~
L54:  'Quit
L55:  'Return to main menu
L56:  '{home}
L57:  '/xmmenu1~
```

continued...

12 Nonparametric Methods

...from previous page

```
I62:  'Menu5 - Test Options
I63:  'Runs
I64:  'Execute the Runs Test for randomness
I65:  '/xcruns~
I66:  '/xmmenu5~
J63:  'Sign
J64:  'Execute the Sign Test
J65:  '/xcsign~
J66:  '/xmmenu5~
K63:  'Mann-Whitney
K64:  'Execute a Mann-Whitney test on pairs
K65:  '/xcmw~
K66:  '/xmmenu5~
L63:  'Quit
L64:  'Return to top Menu
L65:  '/xmmenu1~


I23:  'start
I24:  '{goto}i1~
I25:  '/xmmenu1~


I73:  'Load Subroutine
I75:  '{goto}c22~/fcce{?}~
I76:  '/cc23~c4~/cc25~f4~
I77:  '{home}
I78:  '/xr~
I80:  'Print Subroutine
I82:  '/ppcara20.h20~gll
I83:  'ra1.h20~glllq
I84:  '/xr~
I86:  'Const Subroutine
I88:  0.05
I89:  1.96
I90:  2
I91:  '/xnEnter new test level    ~i88~
I92:  '/xnEnter number of tails   ~i90~
I93:  '/xii88=.05~/xgi98~
I94:  '/xii88=.025~/xgi98~
```

continued...

Nonparametric Methods 12

...from previous page

```
I95:  '/xii88=.01~/xgi98~
I96:  '/xnEnter z from Normal table   ~i89~
I97:  '/xr~
I98:  '{goto}i89~
I99:  '/xii90=1~/xgi103~
I100: '@if(i88=.05,1.96,@if(i88=.025,2.24,@if(i88=.01,2.575,@na)))~
I101: '{edit}{calc}~{home}/xr~
I103: '@if(i88=.05,1.64,@if(i88=.025,1.96,@if(i88=.01,2.33,@na)))~
I104: '{edit}{calc}~{home}/xr~


M73: 'Unload Subroutine
M75: '/rec22.d2047~
M76: '/rec4~
M77: '/ref4~
M78: '/rec9.c11~
M79: '/rec14.c17~
M80: '/ree9~
M81: '/ree12.e15~
M82: '/reg11.g17~
M83: '/xr~


Q1:  'Runs Test
Q3:  '/xnEnter the number of Successes ~c9~
Q4:  '/xnEnter the number of Failures ~c10~
Q5:  '/xnEnter the number of Runs ~c11~
Q6:  '{goto}c14~(1+(2*C9*C10/(C9+C10)))~
Q7:  '{down}(@SQRT((2*C9*C10*(2*C9*C10-C9-C10))/(((C9+C10)^2)*
      (C9+C10-1))))~
Q8:  '{down}((C14-c11)/c15)~
Q9:  '{goto}c17~@if(@abs(c16)c13,1,0)~
Q10: '{home}
Q11: '/xr~
Q14: 'Sign test
Q16: '{goto}e36~
Q17: '/xnEnter probability that sign is pos. ~e9~
Q18: '(1-e9)~{edit}{calc}~/c~e13~/re~
Q19: '@if({left}{left}{left},1,0)~
Q20: '{right}@if({left}{left}{left}{left}{left},1,0)~
```

continued...

12 Nonparametric Methods

...from previous page

```
Q21:  '{left}/c.{right}~.{left}{end}{down}{right}~
Q22:  '{up}@sum({down}.{end}{down})~
Q23:  '/c~{right}~
Q24:  '{goto}e12~(e35+f35)~{edit}{calc}~
Q25:  '{down}{down}
Q26:  '((@max(e35.f35)-(e9*e12))/(@sqrt(e9*e12*e13)))~{edit}{calc}~
Q27:  '/rff4~~{goto}e15~
Q28:  '@if(e14e11,1,0)~
Q29:  '{home}/ree35.f2047~
Q30:  '/xr
Q32:  'Mann-Whitney Test
Q34:  '/xcrs~
Q35:  '{goto}e36~@count({left}{left}.{end}{down})~{edit}{calc}~/c~g11~
Q36:  '@count({left}.{end}{down})~{edit}{calc}~/c~g12~/re~
Q37:  '{goto}g14~
Q38:  '(g11*g12+(g11*(g11+1)/2)-g13)~
Q39:  '{down}(g11*g12*(g11+g12+1)/12)~
Q40:  '{down}(g14-(g11*g12/2))/@sqrt(g15)~
Q41:  '{goto}g17~@if(@abs(g16)g10,1,0)~{home}
Q42:  '/xr~
Q44:  'Rank Sum Subroutine
Q46:  '{goto}d36~
Q47:  '/m{end}{down}~{left}{end}{down}{down}{right}~
Q48:  '{right}/dfe36~1~1~~/df.{left}{end}{down}{end}{down}{right}~~~~
Q49:  '{right}({left}{left}+{left}{left}{left})~/c~.{left}{end}{down}{right}~
Q50:  '{left}{left}{left}/dsrd.{right}{right}{right}{end}{down}{right}~
Q51:  'p{right}{right}{right}{end}{down}~a~g
Q52:  '{right}{right}{right}{right}/dfg36~~~~
Q53:  '/df.{left}{end}{down}{right}~~~~
Q54:  '{goto}c36~
Q55:  '/dsrd.{right}{right}{right}{end}{down}{right}~
Q56:  'p.{right}{right}{end}{down}{left}{left}~d~g
Q57:  '{goto}g35~@sum({down}.
Q58:  '{left}{left}{left}{left}{end}{down}{right}{right}{right}{right})
Q59:  '{edit}{calc}~/c~g13~/re~~
Q60:  '{goto}c36~/dsr
Q61:  'd.{right}{right}{end}{down}~p{right}{right}.{end}{down}~a~g
Q62:  '{right}{right}/re.{right}{right}{end}{down}~
Q63:  '{left}{end}{down}/m{end}{down}~{end}{up}{down}~
Q64:  '{home}/xr~
```

PART IV

USER'S GUIDE

CHAPTER 13

USER'S GUIDE

13 User's Guide

People who read the last chapter of technical books have either reached the end of a long journey or are standing in a bookstore deciding whether to spend $22.95. In either case, this chapter is for you. The material covered in Chapters 6-12 will be treated here as a unit; the templates generated so far will be compiled into a statistical system. If you have worked your way through this book, you will have gained numerous tips from the presentations. If you buy the companion disk and start with this chapter as a user's guide, you will still get your money's worth—a statistics package for under $50.

One reminder to readers: this book emphasizes using 1-2-3 macro language to write statistical templates; it is not a statistics text. Many reference books on beginning statistics are available, and neophytes should be sure to have access to one. This book gives more detail on how to make statistical functions work than on why they work and why they should be used.

A template package is not a formal software package. There are no built-in safeguards against operator error. Certain mistakes can provide erroneous answers, and operators can erase needed portions of the spreadsheet. This is the price you pay to gain the power and flexibility of a template package. All software provided can be reproduced freely, however. If you work only with backup copies, no permanent damage can be done. The master disk (either the one you generate or a purchased companion disk) should be write-protected immediately; then, no problem can occur.

THE COMPOSITE MENU

Seven templates have been constructed and are described thoroughly in Chapters 6-12. Six spreadsheets containing these templates are described in this chapter. Also shown are the NIM and TRANSL templates described in Appendices C and D. Routines are grouped so that each spreadsheet provides a set of related tools. The Menu spreadsheet described in Appendix D allows all the spreadsheets to be run as a cohesive software package. Operation in this mode can easily be handled by data entry personnel.

User's Guide 13

Each chapter contains a figure describing the menu structure for the template it contains. Figure 13.1 is a diagram of the MENU macro tree, covering the major areas of operation of this system. In essence, this is the composite menu listing all the template spreadsheets. The statistical functions available in each template are as follows:

Template	Function
DATAIN	Creates data files for use by all templates
SMALL	One-way analysis of small samples: Min, Max, Avg, Median, Mode, MMA graph
DESCRIPT	One-way analysis of large samples: Mean, Deviation, Variance, Kurtosis, Skew
CHI_SQ	One-way analysis of large samples: Mean, Deviation, Variance, Kurtosis, Skew Freq distribution, Histogram, Chi-Sq fit, Ogive
TWO_WAY	Two-Way analysis: Regression, F test, T test, Z test, scatter plots, smoothing, Gauss plots
NON_PARM	Nonparametric (distribution-independent) methods: Runs test, Sign test, Mann-Whitney U test
NIM	Macro menu demonstration: Play the game of NIM

13 User's Guide

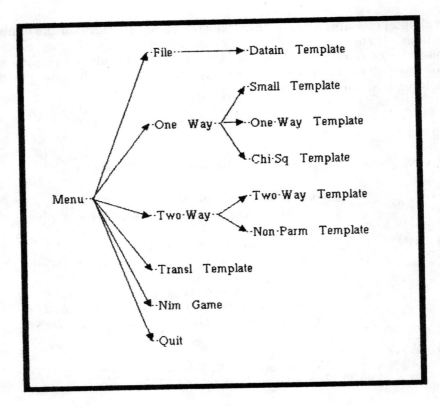

Figure 13.1 Menu Macro Tree

INSTALLING THE PROGRAM

When you receive the companion disk, several things must be done. The first, and most important, is to write-protect it and make a backup copy, which will be your working disk. Instructions for making backup disks are found in Appendix G. If you are manually entering these programs, the same precautions should be taken. Make all program entries onto a working disk or disks. Transfer a completed copy of each spreadsheet onto a master disk. Write-protect this disk and back it up to generate working copies. In either case, store the master diskette with your Lotus 1-2-3 masters.

Tips on Customizing

After the master disks are secure and a working disk is available, you can tailor the package to your environment. Titles in the marquis can be altered, print-out titles can be changed, tests can be altered to specific needs, and, most importantly, the disk drive configuration can be set to the user's needs.

If, for example, XYZ Inc. purchased the companion disk for their quality assurance department, they would want to customize outputs prior to use. The steps that follow can be used to modify the Non__parm template (non-parametric methods). Figure 13.2 shows the marquis resulting from these actions. Each of the other spreadsheets could be upgraded using an identical method.

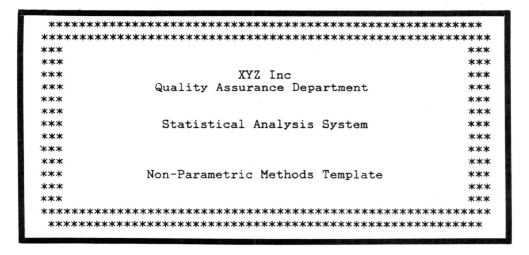

Figure 13.2 Modified Marquis

1. Work only with a backup of the original diskette. Load the file Non__parm. <Esc> halts macro operation.

13   User's Guide

2.   Move the cursor to cell J7. The command line should display the contents as follows:

    ***   STASTICAL MODELING WITH LOTUS 1-2-3   ***

    Using the edit function (), or by reentering the entire line, substitute the following:

    ***   XYZ Inc   ***

3.   In a similar fashion, replace J8 and J11 with the following:

    ***   QUALITY ASSURANCE DEPARTMENT   ***
    ***   STATISTICAL ANALYSIS SYSTEM   ***

4.   Adjust the summary table by changing B2 to the following:

    XYZ Inc Statistical Analysis System

5.   Finally, dress up the printed report with a custom printer command. The commands that follow set the print to expanded mode for two lines in the printed report. The command is shown for an Epson Fx-type printer (\014). If this does not work on your printer, consult the manual or Appendix C in this text.

User's Guide 13

/mI82<return>I81	Move contents of I82 to I81.
<f5>I82<return>	Move cursor to I82.
'os\014~qrI70~grI71~gos{esc}~q	Enter new print commands as a label.
<f5>I83<return>	Move cursor to I83.
'ra2.h20~g111q	Enter modified print range as a label.
<f5>I70<return>	Move cursor to I70.
' XYZ Inc	Enter label line.
<Down>	Move cursor to I71.
'Statistical Analysis report	Enter label line.

By changing only three lines in the spreadsheet and three lines in macros, and by adding two lines, a custom report is generated. The headings can contain anything you want, including special fonts (if your printer supports them). If you have made changes up to this point, generate a new master disk (in addition to the original) and make working backup disks from it.

Disk File Options

The final installation involves setting up your working disk for actual use. The following two sections will instruct you in setting up the system for use on hard disk and floppy disk systems.

Floppy Disk Systems

For a floppy disk system, this system can be set up in two ways. The easiest way is to provide working copies of the template package to users who need them. Created data files are stored on the working disk along with the template. As the disk fills, data files can be transferred to other disks, or additional working copies can be made.

13 User's Guide

If a more structured environment is preferred, with tighter control over the software, the working programs can be transferred to the Lotus 1-2-3 system disk. This procedure requires removal of the COMMAND.COM file from the Lotus disk to provide space. Lotus release 1A occupies approximately 265K on a floppy disk, leaving 95K after the command file is removed. The template package occupies 88K, which leaves plenty of space for customization. Release 2 users must sacrifice the help utility to create space — not necessarily a good choice.

With the template files on the system disk, all that remains is to reroute the directory path. Upon loading Lotus, execute the following:

```
/fda:<return>
```

This command will set the path to the system disk where the templates are located. Remember, 1-2-3 will not accept this change unless there is a disk in drive B.

Each template requires two additional lines to complete this modification. The Autostart macro found at cell I23 must include

```
/fdb:~
```

to set the path to drive B for data file saves. The Quit macro found in Menu1 of each template must include

```
/fda:~
```

to reestablish the path to drive A, where the template files are located.

Hard Disk Systems

The best method for installation on hard disk systems is to copy all templates to the directory in which 1-2-3 resides and make path modifications similar to those described above. Data should be routed to drive B as a default condition. Users can set alternate paths while using the templates.

Boot up the computer and change directories to the one in which Lotus is stored. In most cases, this will be a subdirectory named LOTUS. Insert the working disk in drive A, and type

copy a:*.wks<return>   for Lotus release 1A

or

copy a:*.wk1<return>   for Lotus release 2

All the template spreadsheets will be transferred to the hard disk.

Modify each of the spreadsheets to contain

/fdb:~

in the Autostart macro at cell I23. The Quit macro found in menu1 of each template should contain a path change back to the hard disk. If the files are in subdirectory LOTUS on drive C, the following command should be added:

/fdc:\lotus~

The only exception is the MENU template; the above command should set the path to drive B on exit. If this is not done, 1-2-3 will continue looking for spreadsheet files in the LOTUS subdirectory of the hard disk.

In most cases, one of the floppy drives (usually B) is used for this purpose.

Execution is simple if these instructions have been followed. The user loads 1-2-3 normally and then enters

/fdc:\lotus <return>   Changes path to \lotus
/frmenu <return>   Executes the menu template

393

The Subst (substitute) command described in Appendix G allows a single-character drive name (i.e., F:\) to be substituted for a path name such as C:\LOTUS. In that case, the operator would change the directory to F:\ instead of C:\LOTUS in the previous example.

RUNNING THE TEMPLATE PACKAGE

After Lotus 1-2-3 has been loaded and the path established, the operator can load either the Menu spreadsheet or, directly, one of the application templates. The command line at the top, when MENU is loaded, presents six options: **File**, **One-way**, **Two-way**, **Nim**, **Transl**, and **Quit**.

The following sections describe operations in the various templates.

Data File Creation

Move the cursor to the File option, or press <F> from the opening menu of the MENU template. The file manipulation template (DATAIN) will be loaded. There are seven options presented on the command line when the load is complete: Enter, Retrieve, Import, Trim, Wipe, eXtract, Quit.

When **Enter** is executed, the cursor is moved to the top of the data area, and macro execution is halted. It is possible to provide a macro to input data into a table, but one should not be necessary here. The user is expected to be comfortable in the spreadsheet environment, so he or she can enter the data directly.

Following is a list of variables to be entered. All are stored with the data file.

Variable	Description
Dmnsion	Dimension of the data file (1 = one-way, 2 = two-way)
Name	Name of the data file
Descrip	Description of the data file
Operator	Name of the person responsible for the data
Name	Name of the variable(s) contained in the file
Units	Variable Units

continued...

User's Guide 13

...from previous page

Variable	Description
Hi Limit	Upper limit on data in the file
Lo Limit	Lower limit on data in the file
Count	Number of data elements
Title 1	Graph title first line
Title 2	Graph title second line
X-title	Title for X-axis
Y-title	Title for Y axis
Flt Lbl	Floating label to be used to distinguish between groups, etc. Essentially a short description
Start	Position for the first data value

The user need not input a value into each of these variables. In most cases, not all values will even be available. To make the data file universal, a blank will be stored where no value exists, providing the means to input a data set, along with the related information, and perform several analyses with no additional data entry. The blank also prompts the user to be complete in recording data at an early point in the analysis.

After the informational variables are entered, the actual values—referred to as **raw data values**—are entered. The raw data must start in the cell adjacent to the label START. If the file is dimension 1, the values are a single continuous column of numbers. Zeroes must be entered, and no blanks are allowed; the eXtract utility will save everything up to the first blank in the data set. For dimension 2, data are entered into both columns C and D.

After data entry is complete, the macro must be restarted. <Alt/R> restarts the macro, and the command line contains the same options as were available before.

The second command line option—**Retrieve**—allows a previously saved data file to be reentered back into the template. Files that are similar in nature are often created, and the user has the option of modifying one of these files rather than creating a new one from scratch. The Retrieve option is also useful for correcting data errors.

395

Wipe and **Trim** are options for clearing data from the template. Wipe removes the whole data set (i.e., wipes it out), which is the method for emptying the template to input more data. Trim removes only the raw data. In cases where multiple data sets are to be entered, frequently only the raw data and floating label must change. The Trim option allows removal of the raw data without removal of the data labels for these cases.

Import allows raw data from an ASCII data file to be entered directly into the raw data area. This is a handy feature for transferring data from applications programs into the spreadsheet.

File eXtract is the means for transferring the contents of the data columns directly onto disk. This procedure is the same as saving the data file, but with the eXtract option, only the data values are saved — not the entire spreadsheet.

Quit does not exit 1-2-3; rather, it reloads the original Menu spreadsheet.

Small Sample Techniques

If the One-Way option is selected at the top menu, the user is given three options: SMALL, DESCRIPT, and CHI__SQ. The Small option loads the small sample template. The primary menu in this template offers the user six options: **File**, **Print**, **Graph**, **Cursor**, **Adjust**, and **Quit** .

After selecting **File**, the user can either **Load** or **Unload** a file. The Load macro performs almost all calculations associated with this template. For that reason, no secondary test menu exists. The Unload macro removes the data file and prepares the column for additional entry. This template is the only one set up for multiple-file entry. The user can create a Min-Max-Avg graph with a single command from a multiple data set.

When **Print** is selected, the user can send the summary table to the printer and add either the actual data or a graph below the table as desired. The **Graph** command sends the .PIC file to disk, but, if a graph is to be printed, the standard printgraph utility must be used. This requires that the user exit 1-2-3.

The **Cursor** option allows the user to move around the spreadsheet without stopping the macro. This is especially useful when the calculations for median, mode, and skewness are not within expected ranges. These calculations are performed here with a cellular technique that is sensitive to bin size. The cursor commands allow direct inspection of the scratch pad table where these calculations are made. Based on the results, adjusting the bin size or starting point may be necessary.

The **Adjust** commands allow the values just mentioned to be altered and the data to be recalculated. The **Recalculate** option also provides a method for direct entry of data into this template; it executes the portions of the Load macro that compute the required test values.

Executing **Quit** from the primary menu reloads the menu spreadsheet again, providing the options just discussed.

Large Sample Techniques

The DESCRIPT and CHI_SQ templates are simliar to one another and to the small sample template discussed previously. The same options are presented with slightly altered functions. Only a single data file can be entered into these templates, and the analyses are largely performed in the Load macro, as was the case with small samples.

DESCRIPT

When a file is loaded into the DESCRIPT template, the user is prompted to enter two values. The first value is the top of the lowest cell and should be chosen to ensure that the frequency distribution includes the lowest cell of interest in the analysis. The second value to be entered is the cell size. Two criteria apply to this choice: the cell size should not be less than the resolution of measurement, and the cell size should be large enough to spread all the values in the distribution over the 30 cells that will be created.

13 User's Guide

In an earlier example, the data set was heights of individuals, measured to the nearest inch. If a cell size of 1/2" were selected, every other cell would be empty, so 1" per cell would be the minimum resolution (cell size) that would produce useful output. If the range in heights were 35" to 75", though, only those between 35" and 65" would be included in the calculation (because 30 cells are used). In this case, a 2" cell size should be entered to avoid truncation. The histogram output does indicate the number of points above and below the chart. Allowing truncation to occur is often advantageous, but it should be a conscious decision.

After the cell limits are entered, the calculations are made and displayed on the screen. Included are the following:

- Mean
- Variance (n and n-1 weightings)
- Standard Deviation (n and n-1 weightings)
- Minimum
- Maximum
- Range (max-min)
- Kurtosis
- Skewness
- Count (number of elements in the distribution)

The **Print** option allows output of the results to the printer. Optionally, the user can add a histogram or frequency distribution table to the results.

A single graphic output — a histogram — is available. This output can be substituted for the frequency distribution table cited earlier.

A cursor control menu allows the user to move the cursor around the spreadsheet to view tables and interim calculations without exiting macro operation.

Finally, the **Adjust** menu allows the user to recalculate a spreadsheet after a change is made in the data set. This change can include entering a raw data set from scratch.

Quit returns control to the MENU macro described in Appendix D.

CHI_SQ

The CHI_SQ template is almost identical to the DESCRIPT template. In addition to the calculations performed above, the CHI_SQ template computes a value for a test comparing the data distribution to a normal distribution with the same mean and standard deviation. If the value of the chi-squared test statistic is less than the corresponding value in the table in Appendix A (based on desired accuracy and sample size), the distribution can be considered normal.

The **Print** option has been expanded. In addition to the summary table, several graphs have been added, which can be appended to the summary table. Five different graphic solutions are available. **Histogram** generates a bar chart in a .PIC file. **Data** provides a line plot of the data in the order it appears in the file. **Ogive** provides a cumulative frequency distribution analysis, which is a useful tool for empirical test analysis. Two forms are presented for displaying the **Gaussian** curve based on the calculated mean and standard deviation.

The **Adjust** menu has been expanded to include a sort function. In conjunction with the data plot, the sort function can provide insight into the nature of the data set. The raw data values can be sorted in ascending or descending order. If necessary, they can be returned to their original values by sorting in index order.

Two-way Analysis

When the Two-way option is selected from the menu template, the user can select **Two-Way** (parametric) or **Nonparametric** methods. The parametric methods all require a data set with a dimension of 2. This data can be entered directly, or two single data sets can be merged.

Four statistical tests can be performed by this parametric template. Each must be requested individually. The tests differ enough in nature that executing all of them automatically would hold no advantage. The program is set up to accept a data set and then allow the analyst to run any of these procedures with a few keystrokes.

The **T test** is a small sample comparison test between two means. The **Z test** is similar but applies to larger samples. The **F test** is a test for the significance of the magnitude of the ratio of two variances. **Linear regression** tests the significance of the relationship between two variables. Forecast or **project**ed values can be computed based on the linear regression.

Graphic methods presented allow scatter plots to be made and regression curves to be added. A double Gaussian curve can be plotted to compare the distributions — based on means and deviations — of two samples. Curve-smoothing can be applied to mask random variation due to noise in measurement.

Nonparametric Methods

The three nonparametric methods have been chosen here because they can be applied to a majority of problems requiring this type of analysis.

The **runs test** is applied to a data set to see whether the pattern is random or biased. This knowledge is useful for qualifying empirical data sets for random testing.

The **sign test** is a paired test similar to the T test. The difference between pairs of readings is recorded, and a statistic is calculated based on the relative occurrence of positive and negative differences. The magnitude of the difference is compared to a standard value to predict if the samples are from the same population.

The **Mann-Whitney U test** is the best of the nonparametric methods because its **rank-sum** technique is used to weight the magnitudes of the elements. More information is used here than in the sign test. Two samples of different sizes can be compared, and the hypothesis that they are from the same population can be evaluated. The U test is the only test here that is always two-tailed.

EXAMPLES

Case Study Eight: Vendor Qualification

A manufacturer of shims, which are specified at 9 +/- 5 thousandths of an inch, has shipped five lots to XYZ Inc. Seven samples were measured from each lot, and the results are tabulated in Table 13.1. Find the median, mode, and skewness of each lot. Combine the results of the five lots, and determine whether the population from which they are extracted is Gaussian, based on the chi-squared test statistic. Find the two- and three-sigma limits on this population.

Lot 1: 4, 7, 8, 9, 9, 10, 12
Lot 2: 6, 8, 8, 9, 10, 11, 13
Lot 3: 7, 8, 9, 10, 11, 13, 14
Lot 4: 5, 7, 8, 8, 9, 10, 12
Lot 5: 6, 7, 9, 9, 10, 11, 12

Table 13.1

Solution

The first step in solving this problem is to create a set of data files for the lots in Table 13.1. Load 1-2-3 and retrieve the MENU template, or go directly to DATAIN to create these files. If you are using the companion disk, the following commands should be executed (with a backup copy of the companion disk in drive B):

```
/frmenu<return>         Loads the MENU template
<F>ile                  Loads DATAIN
```

At this time, you will be prompted to choose the (E)nter command (or press <return>):

```
<E>nter                 Begins data enter mode
```

The cursor has moved to cell C10 after printing a message about restarting the macro. Enter the following data, pressing the down arrow key after each entry:

File	Dimension	1
	Name	EX13_1
	Descrip	Example
	Operator	(Operator's name)
Variable	Name	Thickness
	Units	Thousandths
	Low Lim	4
	Hi Lim	14

continued...

...from previous page

	Count	(Leave Blank - program enters value)
Graph	Title 1	Shim Thickness Example
	Title 2	Chapter 13
	X Title	Thickness
	Y Title	Thousandths
	Flt Label	GR1
	Start	4
		7
		8
		9
		9
		10
		12

<Alt/R> restarts the macro. The count is entered, and the main menu is reactivated. At this point, save the data file to floppy disk with the following commands:

e<X>tract	The Save Command
ex13_1	File Name
<T>rim	Remove data elements
<E>nter	Ready for data entry

If you followed the steps above exactly, the seven data elements were removed, leaving the text elements. Following these steps is easier than reentering all the information. Using the Down Arrow key, change the file name cell to ex13__2, and enter the values for lot 2. Press <Alt/R> to restart the macro, and use the eXtract command to save the data to file ex13__2.

Repeat this process for data lots 3-5. After the fifth data block is saved, exit DATAIN and load the SMALL template:

Quit	Exits to MENU template
<O>ne-Way<S>mall	Loads SMALL template
<return>	Response to the "Press Enter" prompt

Now, the menu for the SMALL template is active on the command line. The cursor is at the top of the first column, waiting for a File Enter command:

<F>ile <L>oad ex13_1   Loads file ex13_1 in col 1
0<return>1<return>   Responses to cell size
                                and low cell query

At this point, the results of the calculations for the first data set are displayed on the screen. The mean is 8.428571, and the median is 8.5. To display the remaining four files in the next four columns, the cursor must be moved to the top of column 2:

<C>ursor <R>ight <Q>uit   Moves cursor

The answers to the original questions are summarized in Table 13.2, as extracted from the outputs generated:

	Lot 1	Lot 2	Lot 3	Lot 4	Lot 5
Mean	8.42	9.29	10.29	8.43	9.14
Median	8.5	9	10	8	9
Mode	9	8	14	8	9
Skew (Pears.)	-.03	.35	120	.208	.073

Table 13.2

In the third data set, the mode of 14 may seem incorrect. Actually, there were seven values, each of which occurred once. The bimode indicator confirmed that the mode was not unique. In fact, each of the seven values was the mode — or most common — value. The particular sort used finds the last value and displays it.

If there are any blank columns after column 1, the Graph command will not function properly. Load the second file exactly as you did for file ex13_1, making sure it is adjacent to the first data column. Repeat the cursor move and File Load procedure for the final three files.

13 User's Guide

When all five files are loaded, print a summary report (see Figure 13.3),

<P>rint <S>ummary

and create the Min-Max-Avg graph (see Figure 13.4):

<G>raph <C>reate <S>ave <Q>uit

```
            STATISTICAL MODELLING WITH LOTUS 1-2-3

                  SMALL SAMPLE STATISTICAL ANALYSIS

          GROUP      1         2         3         4         5        6
File              EX13_1    ex13_2    ex13_3    ex13_4    ex13_5    _____
Description       Example   Example   Example   Example   Example   _____
Operator          Casey G   Casey G   Casey G   Casey G   Casey G   _____

Date              01-Feb    01-Feb    01-Feb    01-Feb    01-Feb    _____
Variable          ThicknessThicknessThicknessThicknessThickness    _____
Units             Thous.    Thous.    Thous.    Thous.    Thous.    _____
Low Limit            4         4         4         4         4      _____
High Limit          14        14        14        14        14      _____
Data Count           7         7         7         7         7      _____

Data Label        Gr1       GR2       GR3       GR4       GR5

Minimum              4         6         7         5         6
Maximum             12        13        14        12        12
Average           8.428571  9.285714 10.285714  8.428571  9.142857
% OL Lo            0.0       0.0       0.0       0.0       0.0
% OL Hi            0.0       0.0      14.3       0.0       0.0      _____

Median             8.5        9        10         8         9
Pears Skew        -0.03077  0.134839  0.120385  0.208012  0.072932  _____
Mode                 9         8        14         8         9      _____
Bi-Mode              0         0         1         0         0      _____
```

Figure 13.3 Summary Report

User's Guide **13**

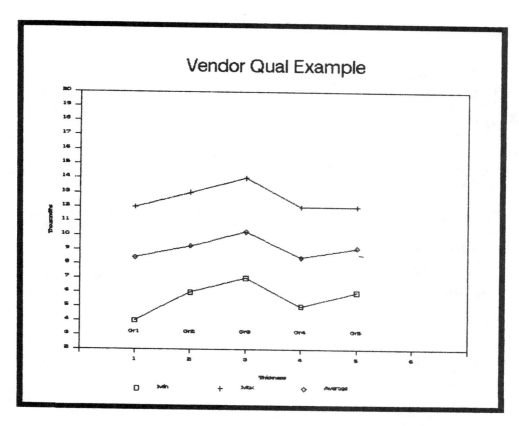

Figure 13.4 Min-Max-Avg Graph

At the main menu, execute Quit to reload the MENU template. Enter <F>ile to reload the DATAIN template. Create a file named ex13__6, combining the five data sets by using the method described in Exercise 6.2 (from Chapter 6) or by entering all of the elements from scratch. E<X>tract this data to file ex13__6, and <Q>uit file mode. Enter the CHI__SQ template by executing the following:

<O>ne-Way <C>hi-Squared

Load the newly created data file:

```
<F>ile<L>oad ex13_6<return>
0<return>1<return>                    (Cell size limits)
<P>rint<S>ummary<C>hi-squared
<Q>uit
```

The results of the chi-squared analysis will be printed in addition to being displayed on the screen.

Based on the output from the CHI__SQ template, the distribution is Gaussian. The value of chi-square — 2.3269 — is well below the critical value of chi-square from the tables (43.77 for 30 D.F. @ .05).

Finally, the mean and unbiased standard deviation are used to find the two and three sigma limits:

Mean = 9.11429
Std Dev = 2.31073
2 Sigma = 4.49283, 13.73575
3 Sigma = 2.1821, 16.04648

Case Study Nine: Marketing

Table 13.3 shows sales for a division of XYZ Inc. for all of 1986. Figures indicate total sales per month in thousands of dollars. Based on these figures, were sales increasing or decreasing at the close of the year? What is the projected sales total for the twelfth month of 1987?

Month	Sales (thousands)
1	375
2	269
3	354
4	322
5	385
6	47
7	404
8	334
9	319
10	301
11	299
12	314

Table 13.3

Solution

Enter the data in Table 13.3 using the DATAIN template. The only difference in entering two-way data is the extra column of data values. Save the data (using eXtract) into a file named ex13__7. Answering the questions for this exercise requires a regression analysis. Enter the TWO-WAY template from the Menu template as follows:

<T>wo-Way<T>wo-Way   Loads TWO__WAY template
<return>   Respond to prompt
<F>ile<L>oad<L>oadex13_7<return>
                                  Loads Data File into
                                  TWO__WAY Template
<T>est<R>egression   Computes regression

On the screen, the regression equation can be seen near the center. In this case, the equation is

$Y = 355.3181 + -3.08741X$

or

13 User's Guide

Y = 355.3 - 3.1X

Since the slope of the regression line is negative, the XYZ company exited 1986 with a negative sales trend. Continue to find the projected value for December 1987 as follows:

```
<P>roject                              Executes projection
24                                     Response to "For an X of"
```

The projection is displayed in the center of the spreadsheet area. The projected Y is 281.2202.

The projected sales amount for December 1987 (the 24th month, based on the regression line) is 281. As expected, it is smaller than the 1986 value because the trend was downward. A regression graph can be created with the following commands:

```
<Q>uit                                 Returns to first menu
<G>raph<C>reate<R>egression            Displays the graph
<Q>uit<Q>uit<Q>uit                     To exit
```

Figure 13.5 shows the regression graph as it was displayed on the screen.

408

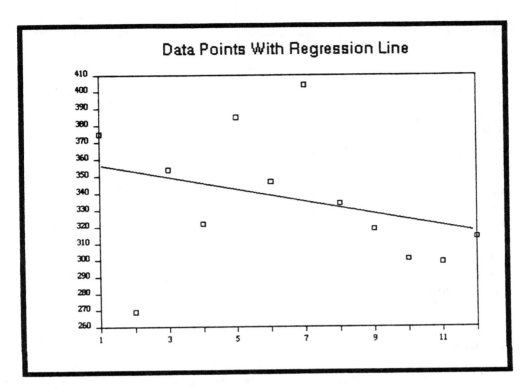

Figure 13.5 Case Study Eight — Regression Graph

Case Study Ten: Engineering

A plastics manufacturer wants to improve its process by making its plastic harder. The data in Table 13.4 represents two runs of plastic — the first with the existing process and the second with the new process. Is the increase in mean significant enough to represent an improvement in the process rather than just random variation? Is the decrease in standard deviation significant? What is the probability that a sample pulled from the new process will be harder than a sample from the old process?

Hardness	Old	New
42		
43	1	
44	1	
45	3	
46	4	1
47	8	2
48	10	7
49	11	9
50	9	13
51	6	10
52	4	7
53	2	5
54	1	4
55		5
56		

Plastic Hardness
Table 13.4

Solution

The data in Table 13.4 must be entered as a two-way file exactly as was done in the previous example. Note that if Gaussian curves are to be plotted later, it is necessary to enter low and high limits for the curve at this time. For this data set, use 40 in column A as a low limit and 60 in column A as a high limit.

Save the entered data sets in a file named ex13__8. There should be 60 values in each column when the count is calculated.

Load the Menu template and execute the following:

<T>wo-Way <T>wo-Way   Loads the TWO__WAY template
<return>   Respond to query
<F>ile <L>oad <L>oadex13_8<return> Loads the data file
<G>raph <C>reate <G>auss   Generates the Gaussian curves
                            Follow instructions on screen to complete and save the graph

<Q>uit

Figure 13.6 shows the intersection of the two curves as they appear on the screen. The right-hand lobe is taller and narrower, with a slight shift to the right in mean. The Z and F tests will indicate whether the shift is significant:

```
<T>est<Z>test
<F>test
1.53<return>                              Response to the "enter
                                          test critical value" prompt
<Q>uit<P>rint<S>ummary<T>est Results      Prints the results
```

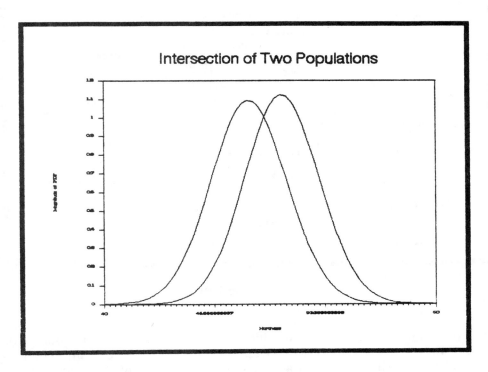

Figure 13.6 Intersection of Plastics Curves

Figure 13.7 shows the screen as it appears after these tests have been executed. The mean in the second group is 50.5, compared to the original mean of 48.75. The Z test rejects the hypothesis that there is no difference in means for the two-tailed test at the 0.05 level. The difference is statistically significant.

```
                STATISTICAL MODELLING WITH LOTUS 1-2-3
                         Two Way Analysis Template

        File: Engineering            Operator: Annette C

                       Sample Statistics
                                Standard Deviation
                     Mean    N Weight   N-1 Wght        Count
        Sample X . . .   48.75  2.270278  2.289437         60
        Sample Y . . .50.74603  2.210855  2.228613         63
        Points of Intersection 123.9496  49.69268

                 Regression Information         Standardized NRV
        Y=           +              X           X1=     32.85
        Standard err of estimate. .             X2=      0.41
        Coefficient of correlation.             Y1=     32.85
        % of Y variation due to X .             Y2=     -0.47
        ****************************************************************
                   T Test    Z Test                      F Test
                  Sml Smpl  Lrg Smpl                     ANOVA
        Test Level . .  0.95    0.95     Test Level . .   0.05
        Crit Value . .          1.96     V1 Value . . .     59
        Deg Freedom. .    74      74     V2 Value . . .     62
        Tails. . . . .     2       2     Crit Value . . . 1.53
        Test Value . .       -4.93655    Test Value . .1.054478
        Hypothesis . .            1      Hypothesis . .      0
          (Acc=0,Rej=1)                    (Acc=0,Rej=1)
```

Figure 13.7 Plastics Test Results

The probability of increased hardness can be found from the points of intersection and the four standardized NRV values in the output. The two curves in Figure 13.6 intersect at 49.34, near the center of the graph, and at 63.14, which is far to the right. Since the area under each of the two curves is unity, the associated probability is the area under the right curve that is above and to the right of the area under the left curve. There is a small area to be deducted, to the right of the intersection at 63.14, but it will be neglected.

X1 and X2 are associated with the left curve and express the intersections in terms of the standard deviation. Similarly, Y1 and Y2 are associated with the right curve. X1 and Y1 are nearly equal and express the right intersection of the two curves. The area under either curve between 6.3 Z and infinity is very near zero and thus will be taken to be zero. The area to the right of the left intersection, 49.34, and under the left curve is

A = 0.5 - .10257 = .39743

The area under the right curve to the right of 49.34 is

A2 = 0.5 + .21904 = .71904

In each case, the associated area is a probability. The probability of the area above the left curve is the difference in these two probabilities:

Af = A2 - A = .71904 - .39743 = .32161

There is a 31% probability that if a sample is drawn from the second distribution, it is higher than it would have been if drawn from the first.

APPENDICES

APPENDIX A: TABLES

z	.00	.01	.02	.03	.04	.05	.06	.07	.08	.09
0.0	.000	.004	.008	.012	.016	.020	.024	.028	.032	.036
0.1	.040	.044	.048	.052	.056	.060	.064	.068	.071	.075
0.2	.079	.083	.087	.091	.095	.099	.103	.106	.110	.114
0.3	.118	.122	.126	.129	.133	.137	.141	.144	.148	.152
0.4	.155	.159	.163	.166	.170	.174	.177	.181	.184	.188
0.5	.192	.195	.199	.202	.205	.209	.212	.216	.219	.222
0.6	.226	.229	.232	.236	.239	.242	.245	.249	.252	.255
0.7	.258	.261	.264	.267	.270	.273	.276	.279	.282	.285
0.8	.288	.291	.294	.297	.300	.302	.305	.308	.311	.313
0.9	.316	.319	.321	.324	.326	.329	.332	.334	.337	.339
1.0	.341	.344	.346	.349	.351	.353	.355	.358	.360	.362
1.1	.364	.367	.369	.371	.373	.375	.377	.379	.381	.383
1.2	.385	.387	.389	.391	.393	.394	.396	.398	.400	.402
1.3	.403	.405	.407	.408	.410	.412	.413	.415	.416	.418
1.4	.419	.421	.422	.424	.425	.427	.428	.429	.431	.432
1.5	.433	.435	.436	.437	.438	.439	.441	.442	.443	.444
1.6	.445	.446	.447	.448	.450	.451	.452	.453	.454	.455
1.7	.455	.456	.457	.458	.459	.460	.461	.462	.463	.463
1.8	.464	.465	.466	.466	.467	.468	.469	.469	.470	.471
1.9	.471	.472	.473	.473	.474	.474	.475	.475	.476	.477
2.0	.477	.478	.478	.479	.479	.480	.480	.481	.481	.482
2.1	.482	.483	.483	.483	.484	.484	.485	.485	.485	.486
2.2	.486	.486	.487	.487	.488	.488	.488	.488	.489	.489
2.3	.489	.490	.490	.490	.490	.491	.491	.491	.491	.492
2.4	.492	.492	.492	.493	.493	.493	.493	.493	.493	.494
2.5	.494	.494	.494	.494	.495	.495	.495	.495	.495	.495
2.6	.495	.496	.496	.496	.496	.496	.496	.496	.496	.496
2.7	.497	.497	.497	.497	.497	.497	.497	.497	.497	.497
2.8	.497	.498	.498	.498	.498	.498	.498	.498	.498	.498
2.9	.498	.498	.498	.498	.498	.498	.499	.499	.499	.499
3.0	.499	.499	.499	.499	.499	.499	.499	.499	.499	.499

Table A.1 Area Under Gaussian Curve

v1\v2	1	2	3	4	5	6	7	8	9	10
1	161	199	215	224	230	234	236	238	240	241
2	18.5	19.0	19.2	19.3	19.3	19.3	19.4	19.4	19.4	19.4
3	10.1	9.55	9.28	9.12	9.01	8.94	8.89	8.85	8.81	8.79
4	7.71	6.94	6.59	6.39	6.26	6.16	6.09	6.04	6.00	5.96
5	6.61	5.79	5.41	5.19	5.05	4.95	4.88	4.82	4.77	4.74
6	5.99	5.14	4.76	4.53	4.39	4.28	4.21	4.15	4.10	4.06
7	5.59	4.74	4.35	4.12	3.97	3.87	3.79	3.73	3.68	3.64
8	5.32	4.46	4.07	3.84	3.69	3.58	3.50	3.44	3.39	3.35
9	5.12	4.26	3.86	3.63	3.48	3.37	3.29	3.23	3.18	3.14
10	4.96	4.10	3.71	3.48	3.33	3.22	3.14	3.07	3.02	2.98
11	4.84	3.98	3.59	3.36	3.20	3.09	3.01	2.95	2.90	2.85
12	4.75	3.89	3.49	3.26	3.11	3.00	2.91	2.85	2.80	2.75
13	4.67	3.81	3.41	3.18	3.03	2.92	2.83	2.77	2.71	2.67
14	4.60	3.74	3.34	3.11	2.96	2.85	2.76	2.70	2.65	2.60
15	4.54	3.68	3.29	3.06	2.90	2.79	2.71	2.64	2.59	2.54
16	4.49	3.63	3.24	3.01	2.85	2.74	2.66	2.59	2.54	2.49
17	4.45	3.59	3.20	2.96	2.81	2.70	2.61	2.55	2.49	2.45
18	4.41	3.55	3.16	2.93	2.77	2.66	2.58	2.51	2.46	2.41
19	4.38	3.52	3.13	2.90	2.74	2.63	2.54	2.48	2.42	2.38
20	4.35	3.49	3.10	2.87	2.71	2.60	2.51	2.45	2.39	2.35
30	4.17	3.32	2.92	2.69	2.53	2.42	2.33	2.27	2.21	2.16
40	4.08	3.23	2.84	2.61	2.45	2.34	2.25	2.18	2.12	2.08
60	4.00	3.15	2.76	2.53	2.37	2.25	2.17	2.10	2.04	1.99
120	3.92	3.07	2.68	2.45	2.29	2.17	2.09	2.02	1.96	1.91
INF	3.84	3.00	2.60	2.37	2.21	2.10	2.01	1.94	1.88	1.83

Table A.2 Distribution of the F Ratio

	15	20	30	40	60	120	INF
1	245	248	250	251	252	253	254
2	19.4	19.4	19.5	19.5	19.5	19.5	19.5
3	8.70	8.66	8.62	8.59	8.57	8.55	8.53
4	5.86	5.80	5.75	5.72	5.69	5.66	5.63
5	4.62	4.56	4.50	4.46	4.43	4.40	4.36
6	3.94	3.87	3.81	3.77	3.74	3.70	3.67
7	3.51	3.44	3.38	3.34	3.30	3.27	3.23
8	3.22	3.15	3.08	3.04	3.01	2.97	2.93
9	3.01	2.94	2.86	2.83	2.79	2.75	2.71
10	2.85	2.77	2.70	2.66	2.62	2.58	2.54
11	2.72	2.65	2.57	2.53	2.49	2.45	2.40
12	2.62	2.54	2.47	2.43	2.38	2.34	2.30
13	2.53	2.46	2.38	2.34	2.30	2.25	2.21
14	2.46	2.39	2.31	2.27	2.22	2.18	2.13
15	2.40	2.33	2.25	2.20	2.16	2.11	2.07
16	2.35	2.28	2.19	2.15	2.11	2.06	2.01
17	2.31	2.23	2.15	2.10	2.06	2.01	1.96
18	2.27	2.19	2.11	2.06	2.02	1.97	1.92
19	2.23	2.16	2.07	2.03	1.98	1.93	1.88
20	2.20	2.12	2.04	1.99	1.95	1.90	1.84
30	2.01	1.93	1.84	1.79	1.74	1.68	1.62
40	1.92	1.84	1.74	1.69	1.64	1.58	1.51
60	1.84	1.75	1.65	1.59	1.53	1.47	1.39
120	1.75	1.66	1.55	1.50	1.43	1.35	1.25
INF	1.67	1.57	1.43	1.39	1.32	1.22	1.00

Table A.2 Distribution of the F Ratio
(continued)

D.F.	.050	.025	.010
1	3.8145	5.0239	6.6349
2	5.9915	7.3778	9.2103
3	7.8147	9.3484	11.345
4	9.4877	11.143	13.277
5	11.071	12.833	15.086
6	12.592	14.449	16.812
7	14.067	16.013	18.475
8	15.507	17.535	20.090
9	16.919	19.023	21.666
10	18.307	20.483	23.209
11	19.675	21.920	24.725
12	21.026	23.337	26.217
13	22.362	24.736	27.688
14	23.685	26.119	29.141
15	24.996	27.488	30.578
16	26.296	28.845	32.000
17	27.587	30.191	34.409
18	28.869	31.526	34.805
19	30.144	32.852	36.191
20	31.401	34.170	37.566

Table A.3 Critical Values of Chi-Square

D.F.	.050	.025	.010
21	32.670	35.479	38.932
22	33.924	36.781	40.289
23	35.173	38.075	41.638
24	36.415	39.364	42.980
25	37.653	40.647	44.314
26	38.885	41.923	45.642
27	40.113	43.194	46.963
28	41.337	44.461	48.278
29	42.557	45.722	49.588
30	43.773	46.979	50.892
40	55.759	59.342	63.691
50	67.505	71.420	76.154
60	79.082	82.298	88.379
70	90.531	95.023	100.42
80	101.88	106.63	112.33
90	131.14	118.14	124.11
100	124.34	129.56	135.81

Table A.3 Critical Values of Chi-Square
(continued)

D.F.	N	.1	.05	.02	.01
1	2	6.314	12.706	31.821	63.65
2	3	2.910	4.303	6.965	9.925
3	4	2.353	3.182	4.541	5.841
4	5	2.132	2.776	3.747	4.604
5	6	2.015	2.571	3.365	4.032
6	7	1.943	2.447	3.143	3.707
7	8	1.895	2.365	2.998	3.499
8	9	1.860	2.306	2.893	3.355
9	10	1.833	2.262	2.821	3.250
10	11	1.812	2.228	2.764	3.169
11	12	1.796	2.201	2.718	3.106
12	13	1.782	2.179	2.681	3.055
13	14	1.771	2.160	2.650	3.012
14	15	1.761	2.145	2.624	2.977
15	16	1.753	2.131	2.602	2.947
16	17	1.746	2.120	2.583	2.921
17	18	1.740	2.110	2.567	2.898
18	19	1.734	2.101	2.552	2.878
19	20	1.729	2.093	2.539	2.861
20	21	1.725	2.086	2.528	2.845
21	22	1.721	2.080	2.518	2.831
22	23	1.717	2.074	2.508	2.819
23	24	1.714	2.069	2.500	2.807
24	25	1.711	2.064	2.492	2.797
25	26	1.708	2.060	2.485	2.787
26	27	1.706	2.056	2.479	2.779
27	28	1.703	2.052	2.473	2.771
28	29	1.701	2.048	2.467	2.763
29	30	1.699	2.045	2.462	2.756
30	31	1.697	2.042	2.457	2.750
40	41	1.684	2.021	2.423	2.704
60	61	1.671	2.000	2.390	2.660
120	121	1.658	1.980	2.358	2.617
Inf	Inf	1.645	1.960	2.326	2.576

Two Tailed Values of T

Table A.4 Students' T Distribution

APPENDIX B: ENHANCING OUTPUT WITH PRINTER OPTIONS

Optional print commands include (but are not limited to) the following:

- different fonts with larger, smaller, or different style type

- carriage controls such as tab, line feeds, reverse tabs/line feeds, and form feeds

- print modifications, including bold, enhanced, multi-strike, superscript, subscript, and underline

These optional commands can be used to dress up outputs from the templates in this book or any spreadsheet.

Just as there are new PC models released every year, there are new options for output. Lotus supports a wide range of printers and output devices and even allows customized use of 1-2-3 to support individual options. Figure B.1 has the distinction of being printed with the short macro at the end of this section.

There are five lines in the print demonstration in Figure B.1. The first line is standard, unmodified, dot matrix print. The second line is compressed print (17 characters per inch); the third line is elite (12 cpi), and the fourth line is pica (10 cpi). The final line is expanded print (5 cpi). Look closely and you will see that the standard print is identical to pica.

The next four lines in Figure B.1 repeat the previous print samples with italics mode set. This output is followed by three lines printed with italics reset and both double-strike and emphasized print set. Double-strike print causes the line to be printed twice; emphasized print causes each dot to be printed twice with the second strike moved slightly to the right. The net result is a bold appearance and filled spaces between dots. Compressed print cannot be emphasized because of the small size of the individual letters.

Appendix B: Enhancing Output with Printer Options

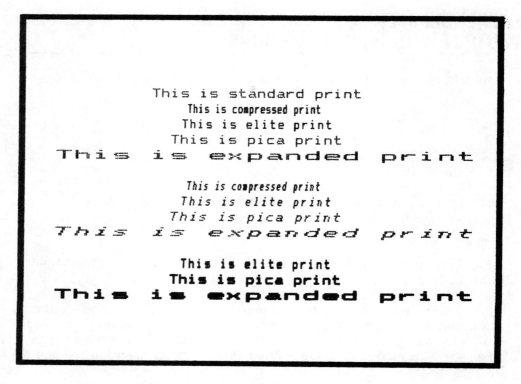

Figure B.1 Printer Code Examples

The following section on print options explains the kinds of features you can expect to use with customized printing. The commands section explains how to find the printer control codes to accomplish these tasks. The setup strings section shows how to send single print modifications to the printer. The final section, macro control of printer options, shows how to send the control codes from a macro. Figure B.1 was created with a single execution of the macro to follow.

If you read this section, you will find how easy it is to make bold or enlarged headings for your spreadsheets, change fonts, or exercise any other print options you want.

Print Options

While a wide variety of options exists, the few demonstrated here were selected for their adaptability and because they are common to many printers. They provide sufficient variation to allow headings to be printed, key words and lines to be emphasized, and variety in type style to be used. Large spreadsheets can be compressed, and, with some printers, short fonts can be used to add lines to a printed page.

Print options can be used in combinations. The nominal compressed and expanded fonts are actually pica compressed and pica expanded. If elite print is expanded or compressed, a different font size results. You can even compress pica print and then expand the result. Table B.1 summarizes the possible combinations and gives the character width, characters per inch, characters per line, and number of columns that can be printed on a page.

Setting	Width	CPI	CPL	Columns
Pica	.083	10	80	8
Compressed (pica)	.041	17	137*	14
Expanded (pica)	.166	5	40	3
Compressed Expanded	.082	8.5	68	6
Elite	.055	12	96	9
Elite compressed	.041	20	160	17
Elite expanded	.110	6	48	4

*Requires resetting of margin

Table B.1
Epson FX-85 Printer Font Sizes

Compressed pica and compressed elite both have the same character size, but the spacing is different. If the elite print size is used, the output from Lotus release 2 can mirror the screen when the 90 × 38 screen display option has been executed at installation. Remember, if the number of columns is increased at the printer, you must increase the right margin in 1-2-3 by executing

/PPOMR   (<p>rint <p>rinter <o>ptions <m>argin <r>ight)

Appendix B: Enhancing Output with Printer Options

The output will then be adjusted to take advantage of that extra capability.

Table B.1 is by no means a complete set of printer outputs, but it shows the options common to almost all currently available dot matrix printers. The FX-85 has a near letter-quality mode that can be selected from 1-2-3. Many printers will print superscripts and subscripts, and many printers have special fonts. In a printer with an active download character set loaded from an external font program, 1-2-3 can switch between fonts on command. The only limit is the user's imagination.

Printer Commands

Table B.2 is an excerpt from the Epson FX-85 command set. Most of the commands shown were used in generating the example shown in Figure B.1. Before you continue this section, find the corresponding commands in your own printer manual and record them in Table B.3. The best place to look for the commands would be a in summary section, appendix, or in a quick-reference card. Substitute them for the commands shown here if you want to run this demonstration program. There may be slight variations in the names for commands. If you are uncertain, experiment with the commands you find. You cannot hurt the printer, the computer, or 1-2-3 by doing so.

Function	On	Off
Compressed	Chr$(15)	Chr$(18)
Expanded (one line)	Chr$(14)	Chr$(20)
Expanded	Esc"W1"	Esc"W0"
Elite	Esc"M"	
Pica	Esc"P"	
Emphasized	Esc"E"	Esc"F"
Superscript	Esc"S0"	Esc"T"
Subscript	Esc"S1"	Esc"T"
Underline	Esc"-1"	Esc"-0"
Italics	Esc"4"	Esc"5"

continued...

...from previous page

Function	On	Off
Line Feed	Chr$(10)	
Form Feed	Chr$(140)	
Reverse Line Feed	Esc"j"	

**Table B.2
Demonstration Printer Control Codes**

Function	On	Off
Compressed	_____	_____
Expanded (one line)	_____	_____
Expanded	_____	_____
Elite	_____	
Pica	_____	
Emphasized	_____	_____
Superscript	_____	_____
Subscript	_____	_____
Underline	_____	_____
Italics	_____	_____
Line Feed	_____	
Form Feed	_____	
Reverse Line Feed	_____	

**User's Printer Control Codes
Table B.3**

There are two types of control codes shown, and they are representative of printers in general. The first type is control characters, such as Chr$(15), and the second type is escape sequences, (e.g., Esc"W1"). Both are explained here.

Appendix B: Enhancing Output with Printer Options

Control codes are usually a single ASCII character sent to the printer. They are generally one of the nonprintable characters at the low end of the ASCII table, which means that they are the same as <Ctrl/Key> combinations. The first example was Chr$(15), used to turn on compressed print. This is a <Ctrl/O> command. The Chr$(n) representation stems from the common use of these commands from BASIC and the fact that most computer users understand BASIC programming.

If a BASIC program LPRINTs an ASCII string, that string is sent directly to the printer. The ASCII table representation for an uppercase A is 65, so if the user executes

```
LPRINT Chr$(65)
```

in a BASIC program, an uppercase A is printed. The command is identical to

```
LPRINT "A"
```

By the same token, if the user executes

```
LPRINT Chr$(15)
```

a <Ctrl/O> command is sent to the printer. Until the command is reset, the type will be compressed. Lotus has provided the setup string utility to allow these codes to be sent to the printer.

Escape sequences work in a similar manner but require more than a single string to be sent. The first character in each of these sequences is the Escape command, or Chr$(27). The Escape command alerts the printer that one or more characters follow for printer control. Expanded mode is set by entering

```
Esc"W1"
```

The "W1" sends an uppercase W and a 1 to the printer, which means "send these ASCII characters," so the proper print command in the BASIC format is

```
LPRINT Chr$(27);Chr$(87);Chr$(49)
```

From this point, the print is expanded until the command is reset. Resetting can be accomplished in two ways. Executing

```
LPRINT Chr$(27)"@"
```

or simply turning the printer off and then back on will do it. Note that everything is reset, including the top-of-form and tab settings.

The easiest way to fully understand these concepts is to try them from a BASIC program. Even if you are not comfortable with programming, you can try this exercise. Follow these steps exactly:

1. Put a disk with BASIC into drive A of your computer. If you are not certain BASIC is on the disk, enter

    ```
    dir bas*.*<return>
    ```

    You should see one or more files such as the following:

    ```
    BASIC.COM
    BASICA.COM
    ```

2. At the system prompt (A), enter

    ```
    basica<return>
    ```

If you are using a MS-DOS-compatible substitute, enter

```
GWBASIC<return>
```

3. When BASIC loads, there will be some information at the top of the screen describing revision level and memory. The last word to appear on the screen should be the BASIC prompt

```
READY
```

4. Type the following lines. If you make a mistake, retype the line entirely. It does not matter if you enter a line out of order; as long as it begins with the correct number, BASIC will reorder it correctly:

```
10 LPRINT "STANDARD PRINT"
20 LPRINT CHR$(15);"COMPRESSED PRINT"
30 LPRINT CHR$(18);CHR$(14);"EXPANDED PRINT"
40 LPRINT CHR$(27);"E";"EMPHASIZED PRINT"
50 LPRINT CHR$(27);CHR$(70);"STANDARD PRINT"
```

5. Then, enter

```
list<return>
```

and the five lines of the program will be displayed on the screen.

6. Enter

```
run<return>
```

and the results will appear on the line printer.

7. Try different commands from the list you generated in Table B.3, or experiment with the ones in Table B.2 if you do not have access to a manual for your printer. Many of these commands will work on printers other than Epson printers.

Setup Strings

If you worked through the last section, you should be comfortable with the way the printer's control codes actually work. In line 20 of the example, the <Ctrl/O> command [Chr$(15)] was sent to the printer before the "compressed print" message in the same line. Lotus has provided the option of doing the same thing using a setup string.

Executing the print menu by entering the following

/pp   (<p>rint <p>rinter)

provides the user with several options. One of these options is

os   (<o>ptions <s>etup)

You are allowed to enter a setup string of 39 characters to the printer. You can send the <Ctrl/O> before printing the message, just as you did in line 20. The steps that follow allow identical output to the BASIC example. This is not a macro. Use the keyboard to enter the contents of the left column exactly as shown.

Appendix B: Enhancing Output with Printer Options

Load Lotus 1-2-3 and execute the following:

<F5>A1<return>	Goto cell A1
STANDARD PRINT<return>	Enter the message into A1
/ppra1<return>gq	Print the first line
COMPRESSED PRINT<return>	Change the message
/ppos\015<return>qgq	Enter the setup string
	Print the second line
EXPANDED PRINT	Change the message
/ppos<esc>\018\014<return>qgq	Enter both setup strings
	Print the third line
EMPHASIZED PRINT	Change the message
/ppos<exc>\027E<return>qgq	Print the fourth line
STANDARD PRINT	Change the message
/ppos\027\070qgq	Reset the printer
	Print the final line

After execution of these statements, there will be five lines on the printer, but they will not be exactly like those in the BASIC example. The first, fourth, and fifth lines all start in the same column. The compressed line starts to the left, and the expanded line to the right.

Lotus automatically indents the amount of spaces specified in the options margins section. The default is four columns. Each of these five lines starts in the fourth column, but the size of the space is the same as the size of a character. Remember in nonproportional dot matrix printing, all characters are the same width, including spaces.

Macro Control of Printer Options

The operations just explained can easily be executed from a macro. Some startling results can be created this way. The following macro generates Figure B.1 (without the border box). The first group of statements contains the labels that are printed. Notice that each line has a different number of blanks preceding the text. As explained previously, this is to compensate for the difference in width of the leading spaces. The information in Table B.1 allows you to calculate the necessary number of spaces, but trial and error is easier. You can make a printout with your first guess and adjust the lines to fit after a test print.

Labels for Print Example

```
A1:  ' This is standard print
B1:  '
A2:  ' This is compressed print
A3:  ' This is elite print
A4:  ' This is pica print
A5:  ' This is expanded print
C7:  ' Printer Code Examples
D8:  ' Figure B.1
```

Line A16 starts the print execution, and the macro is one long print statement. In A17, the letters "ca" entered at the beginning of the line "clear all" of the print ranges. This is done because the "os" command must be reset before a new setup string is added, or the string will become increasingly longer. Lines A17 to A20 print the first group of lines.

```
A15:    'Print Macro
A16:    '/ppacara1.a1~g
A17:    'caos\015~qra2.a2~g
A18:    'caos\018\027M~qra3.a3~g
A19:    'caos\027P~qra4.a4~g
A20:    'caos\014~qra5~g
```

A21 sets the print to italic and prints the blank range A6. A22 through A25 reprint the lines in italics.

```
A21:    'caos\0274~qra6~g
A22:    'caos\015~qra2.a2~g
A23:    'caos\018\027M~qra3.a3~g
A24:    'caos\027P~qra4.a4~g
A25:    'caos\014~qra5~g
```

A26 resets the italics and sets both emphasized and double-strike modes, again printing a blank dummy range to send the string. A27 through A29 print the lines a third time. The first line has been omitted because compressed print cannot be emphasized (because of the printer — not because of 1-2-3).

Appendix B: Enhancing Output with Printer Options

```
A26:    'caos\027@\027E\027G~qra6~g
A27:    'caos\018\027M~qra3.a3~g
A28:    'caos\027P~qra4.a4~g
A29:    'caos\014~qra5~g
```

A30 resets the print to nominal for the printing of the last two lines labeling the picture.

```
A30:    'caos\027@~qra1~g11
A31:    'cara7.h8~g
A32:    'caos\140~qra6~gq
```

Don't be discouraged if you make errors when you first implement these operations into macros. Just use the step function and walk through the operations. Remember, there are no self-destruct commands, so if you are uncertain, experiment.

APPENDIX C:
WRITING SOFTWARE FOR IMPORTING DATA INTO LOTUS

There are many reasons why a user might want to import data into Lotus. If information is gathered or processed by a computer and stored in data files, it may be easier to convert these files into Lotus format than enter the data from scratch. Using a file conversion utility also minimizes errors associated with manual entry of data.

Lotus has provided translation utilities for converting data from several commercial software packages (dBASE, VisiCalc, etc.) and the IMPORT utility for transferring ASCII information into spreadsheet format. This section will describe how to create data files that can be imported into Lotus 1-2-3 and what to do with them when they are imported.

Two listings are provided at the end of this appendix. The first is a BASIC program that will generate a data file that can be translated into the same format as the files created by the DATAIN macro template. This program can be entered into BASICA and executed or compiled using Microsoft QuickBASIC. The program serves two useful purposes. It provides an example of how software can be used to generate ASCII data files for interaction with Lotus, and it also provides a useful utility for users of the templates provided in this text.

The second listing is a macro to be added to the statistical templates already described. Its function is to convert the BASIC program's output into the proper form in Lotus and generate the actual data file, and this listing includes the instructions for modifying the MENU template to call it.

BASIC was chosen because of its universal appeal. Most business majors have had at least one course in BASIC, and all technical majors should have included it. Virtually all programmers in the other high-level languages have enough experience with BASIC to extract the necessary skills from this section.

Appendix C: Writing Software for Importing Data into Lotus

Basic Program Description

The following program is divided into sections in a block format. There is an eleven-line program with seven subroutines, each with a specific function. Line 50 of the main body calls the subroutine that loads all the constants for the program. Line 60 executes the menu subroutine and returns the user selection. Lines 70-90 direct program flow based on this selection. Line 100 returns control to the menu again.

Menu and Constants Subroutine

The Menu subroutine offers the user the following options:

- input data from the keyboard
- input data from a file
- edit/correct data
- print data
- store data
- change directory path
- erase data
- exit

Lines 120-250 print these options on the screen. In line 260, the user selection is entered and is returned in the variable O. Lines 70-90 test values of O to direct program flow.

The subroutine starting at line 1,360 contains 15 variable assignments to set the values for constants used as labels in the program and to set the initial direction for the path command.

Keyboard Entry

The Keyboard Input subroutine is called when either a 1 or 3 is selected at the menu. Line 350 causes the program to loop twice to enter two columns of data. Line 370 aborts the second pass through for data sets with only one column (dimension = 1). This action is necessary because of the way BASIC prints data to the file.

Appendix C: Writing Software for Importing Data into Lotus

If data were entered in pairs and printed to the file with the pairs separated by a comma (or semi-colon), when 1-2-3 imported the pair, it would appear as a single string with a comma imbedded in it. This string would be contained entirely in column A. The Parse command in release 2 could be used to separate these fields, but they would have to be created with the same number of characters. It is easier to send all of the data to 1-2-3 in a single column and use a macro to move that data into the proper positions, which is the function of the macro shown later in this chapter.

Lines 390 to 450 cause the program to loop through the 14 informational data values stored at the top of each data file. Array L$(n) contains the variable name displayed by line 430. The current value in the data array is displayed next, a$(J,I). Line 440 allows the user to enter a blank or a value. If a blank is entered, the data value is left unchanged; otherwise, the new value replaces the old. On the initial keyboard entry pass, all data values will be blank.

Lines 400-420 in this subroutine cause the program to skip the first, second, and fourth data values in the second pass through a two-dimensional array. There is only one value allowed for the file name, operator, or dimension. Lines 460-520 allow entry of the raw data values in exactly the same manner. Printing chr$(7) in line 520 causes the computer to send a beep to the operator when the last value is entered.

File Input

The File Entry subroutine starting in 550 allows the user to recall a previously created data file so it can be edited. This routine will only input data from files created by this utility. Lotus files can be converted to this format, but no benefit would be gained, so that function was not included.

Lines 600-640 allow entry of the file name, addition of the default path, and opening of the file in line 640. The first column of data is read into the data array by lines 650-700. In line 710, the program checks the dimension. If it is 1, the routine halts; if the dimension is 2, dimensional data lines 720-770 read in the second column.

Print Data Subroutine

Line 870 determines which of two paths is taken for printing of the 14 lines at the top of the file. One-dimensional files are lines 880 to 940. Two-dimensional files are lines 950 to 1040. In the raw data section, lines 1010-1020 allow the program to print two-dimensional data files in which the raw data portions have different numbers of elements.

Data Store and Path Subroutines

The Data Store routine is similar to the Input subroutine. The 14 data elements for column 1 are stored, followed by the raw data. If the dimension is two, the second column is printed to the file; otherwise, the routine is ended.

The path routine allows updating of variable P$, which contains the path where files will be stored and recalled.

Macro Description

The following macro will translate a two-dimensional file into the proper format for template use. One-dimensional files need not be translated.

Line 123 of this macro sets the screen display to the marquis. There is no menu because this is a single-purpose template. The screen display prompts the user for the name of a file to translate, converts that name into the proper format, and stores it in a spreadsheet file with the name entered into the data file. For this reason, it is imperative that a file name be entered when the file entry program is run.

A careful examination of the macro reveals that three lines are missing. The execution of a macro halts the first time a blank is encountered, so, obviously, something must be entered. Cell I23 requests the name of a file to import. The label command transfers this name to cell I25, eliminating the first blank. Line I24 copies the same name to line I27, then executes a file import text command. The file name inserted into I25 is entered, followed by the tilde (~) in I26.

```
I23:    '{home}{goto}i1~/xlEnter name of file to import~i25~
I24:    '{home}/ci25~i27~/fit
I26:    '~{right}{right}/fin
```

Line I26 continues with two right arrow commands and another file import. The second import enters values as numbers. I27 contains the same data file name, which is entered by the tilde in I28. I28 continues by transferring the output file name from A2 to I45 with the /c command. The final command in I28 copies the contents of the current cursor location (using /c~) to the cell two spaces to its left. Cell I26 had moved the cursor to C1, so the numerical dimension value in C1 overwrites the string value (both have the same number) in A1. The cursor then moves to C7.

```
I28:    '~/ca2~i45~/c~{left}{left}~{goto}c7~
```

Cells I29 through I36 repeat this process for all cells in the file. You must have a mixture of numerical and text values in the final output file. Since 1-2-3 will not mix them on input, this process effectively produces a single data column with the proper blend.

```
I29:    '/c{down}{down}~{left}{left}~
I30:    '{goto}c15~/c{end}{down}~{left}{left}~
I31:    '{end}{down}~{down}{down}{down}
I32:    '{down}{down}{down}{down}
I33:    '/c{down}{down}~{left}{left}~
I34:    '{down}{down}{down}{down}
I35:    '{down}{down}{down}{down}
I36:    '/c{end}{down}~{left}{left}~
```

Appendix C: Writing Software for Importing Data into Lotus

Line I37 moves to A15, which is the top of the raw data portion of column 1. The {end}{down} sequence then moves the cursor to the last element in column A. The first {down} in I38 moves the cursor to the first element in what will be column B. The /m. sequence anchors the pointer for the move of this column. Fourteen {down} commands are executed to carry the cursor through the data variables at the top of the file. From that point, the {end}{down} in I42 completes the move range entry. Cell B1 is entered as the move to range, followed by a cursor {home}.

```
I37:  '{goto}a15~{end}{down}~
I38:  '{down}/m.{down}{down}
I39:  '{down}{down}{down}{down}
I40:  '{down}{down}{down}{down}
I41:  '{down}{down}{down}{down}
I42:  '{end}{down}~b1~{home}
```

Line I43 erases the numbers column. This step is actually unnecessary, but it makes the screen easier to read. Lines I44-I46, which include the file name entered above, complete the file extract. The final statement reloads the menu template.

```
I43:  '/rec1.d2047~
I44:  '/fxf
I46:  '~a1.b2047~r~
I47:  '/frmenu~
```

Installing the Macro

Two tasks remain for installation of this macro. Enter all the statements from the composite listing at the end of this chapter. Save all the information in a spreadsheet named TRANSL. Enter the following command prior to saving this file:

/rnc\0<return>i23<return>

This sequence initializes the autostart macro.

Appendix C: Writing Software for Importing Data into Lotus

For the Menu template to call this routine, it must be modified. Load the menu using /fr (file retrieve). Press <Esc> to halt the macro. Enter the following:

```
/md33.f36<return>c33<return>         Open a spot in the menu
<f5>f34<return>                      GoTo f34.
Transl<return>
<down arrow>
Translate External Data File<return>
<down arrow>
'/frtransl~<return>
/fsmenu<return>r<return>             Save the corrected file.
```

Using the BASIC Program

The following program can be used for data file generation for templates presented in this book. When BASICA is loaded with the DATAIN file, the operator enters "run" to start execution. From that point, all operations are menu-driven.

To enter a data file from scratch, the operator enters 1 at the menu. The program allows entry of either one- or two-dimensional programs, prompting the user for values as they are needed. The software will check the data count in line 9 and prompt for that many data values to be entered. This may seem like an inconvenience, but the double-check is worthwhile. Have the operator count the values prior to entry and enter that number on line nine. The file name and dimension are also mandatory entries when using the software.

If corrections are required, enter 3 (option 3) at the main menu. The same keyboard entry routine is entered, but the current data value is displayed on each line as data are entered. To keep this value the same, press <return>. To change the value, merely enter the new value and then press <return>.

A data file can be recalled for modification. Option 2 can be used to make a new data file from one with similar settings. For most data corrections, the actual 1-2-3 template will be used.

Option 4 allows a data file to be printed. This feature is handy because an operator can enter data for you and provide you with a hard copy of the file as it is entered. It is also a good way to generate a hard copy for double-checking the data entry.

Option 5 sends the data to the file name entered by the user. Modifying this program to use the file name described in the data set would be a simple procedure. That possibility will be left to the reader as an exercise.

The final option allows entry of a new value for the variable P$, which is combined with each file name to set the data file path. It is best to use path B:\ so the files created are on the 1-2-3 data disk.

After a data file is created, load 1-2-3 and enter the menu spreadsheet. Select the TRANSL utility. The translation macro described here is loaded. Enter the name of the data file created. Nothing else need be done. The final data file will be created with the name listed in the data file itself. The menu program will be reloaded. Run TRANSL again for multiple files. Remember that one-dimensional files do not need to be translated.

BASIC Program Listing

The following program was written in IBM BASICA but should be compatible with any Microsoft version. It has been tested on BASIC and BASICA (both IBM and Compaq) and GWBASIC. It compiles and runs with no errors in Microsoft QuickBASIC.

If you are going to enter this program, load BASIC and type the statements exactly as they appear. Save them in a disk file ("DATAIN.BAS" was used when the program was created) for execution. The B drive was used as the default storage drive so files would be ready to load into Lotus without having to move them.

A copy of this program is included on the companion disk.

Main Program Body

```
10 REM
20 REM DATA ENTRY PROGRAM - GENERATES LOTUS COMPATOBLE FILES
30 REM
40 DIM L$(15),D(2,1000),A$(2,15)
50 GOSUB 1370
60 GOSUB 130
70 IF O=9 THEN STOP
80 IF O=7 THEN RUN
90 ON O GOSUB 300,570,300,820,1070,1290
100 GOTO 60
110 END
```

Menu Subroutine

```
120 REM
130 REM MENU
140 REM
150 CLS:PRINT:PRINT:PRINT:PRINT
160 PRINT TAB(25);"STATISTICS DATA ENTRY PROGRAM"
170 PRINT TAB(30);"CURRENT PATH IS ";P$:PRINT
180 PRINT TAB(24);"1. INPUT DATA FROM THE KEYBOARD"
190 PRINT TAB(24);"2. INPUT DATA FROM A FILE"
200 PRINT TAB(24);"3. EDIT/CORRECT DATA"
210 PRINT TAB(24);"4. PRINT DATA"
220 PRINT TAB(24);"5. STORE DATA"
230 PRINT TAB(24);"6. CHANGE DIRECTORY PATH"
240 PRINT TAB(24);"7. ERASE DATA"
250 PRINT TAB(24);"9. EXIT"
260 PRINT:PRINT TAB(24);:INPUT "ENTER OPERATION NUMBER";O
270 RETURN
```

Keyboard Input/Edit Subroutine

```
280 REM
290 REM
300 REM KEYBOARD INPUT SUBROUTINE
310 REM
320 REM
330 CLS:PRINT:PRINT:PRINT:PRINT
340 PRINT TAB(24);"ENTER OR EDIT DATA":PRINT
350 FOR J=1 TO 2
360 IF J=1 THEN PRINT TAB(24);"ENTER FIRST COLUMN DATA"
370 IF J=2 AND VAL(A$(1,1))=1 THEN 530
380 IF J=2 THEN PRINT TAB(24);"ENTER SECOND COLUMN DATA"
390 FOR I=1 TO 14
400 IF J=2 AND I=1 THEN 450
410 IF J=2 AND I=2 THEN 450
420 IF J=2 AND I=4 THEN 450
430 PRINT L$(I);TAB(18);A$(J,I);TAB(50);
440 INPUT T$:IF T$="" THEN 450 ELSE A$(J,I)=T$
450 NEXT I
460 PRINT:PRINT TAB(24);"ENTER RAW DATA VALUES FOR COLUMN ";J:PRINT
470 FOR I=1 TO VAL(A$(J,9))
480 PRINT I;TAB(6);D(J,I);TAB(18);
490 INPUT T$
500 IF T$="" THEN 510 ELSE D(J,I)=VAL(T$)
510 NEXT I
520 PRINT:PRINT CHR$(7)
530 NEXT J
540 RETURN
```

File Input Subroutine

```
550 REM
560 REM
570 REM FILE INPUT ROUTINE
580 REM
590 REM
600 CLS:PRINT:PRINT:PRINT:PRINT
610 PRINT TAB(24);"ENTER NAME OF INPUT FILE"
620 PRINT TAB(24);"INCLUDE EXTENSION";
630 INPUT F$:F$=P$+F$
640 OPEN"I",1,F$
650 FOR I=1 TO 14
660 INPUT#1,A$(1,I)
670 NEXT I
680 FOR I=1 TO VAL(A$(1,9))
690 INPUT#1,D(1,I)
700 NEXT I
710 IF VAL(A$(1,1))=1 THEN 780
720 FOR I=1 TO 14
730 INPUT#1,A$(2,I)
740 NEXT I
750 FOR I=1 TO VAL(A$(2,9))
760 INPUT#1,D$(2,I)
770 NEXT I
780 CLOSE#1
790 RETURN
```

Print File Subroutine

```
800 REM
810 REM
820 REM PRINT DATA ROUTINE
830 REM
840 REM
850 LPRINT:LPRINT:LPRINT
860 LPRINT TAB(20);"DATA FILE CONTENTS":LPRINT
870 IF VAL(A$(1,1))=1 THEN 880 ELSE 950
880 FOR I=1 TO 14
890 LPRINT L$(I);TAB(20);A$(1,I)
900 NEXT I
910 FOR I=1 TO VAL(A$(1,9))
920 LPRINT TAB(20);D(1,I)
930 NEXT I
940 RETURN
950 FOR I=1 TO 14
960 LPRINT L$(I);TAB(20);A$(1,I);TAB(45);A$(2,I)
970 NEXT I
980 M=VAL(A$(1,9)):IF VAL(A$(2,9))>M THEN M=VAL(A$(2,9))
990 PRINT
1000 FOR I=1 TO M
1010 IF I<=AL(A$(1,9)) THEN LPRINT TAB(20);D(1,I);
1020 IF I<=AL(A$(2,9)) THEN LPRINT TAB(45);D(2,I
1030 NEXT I
1040 RETURN
```

Create Data File Subroutine

```
1050 REM
1060 REM
1070 REM STORE DATA ROUTINE
1080 REM
1090 REM
1100 F$=P$+A$(1,2)+".PRN"
1110 OPEN"O",1,F$
1120 FOR I=1 TO 14
1130 PRINT#1,A$(1,I)
1140 NEXT I
1150 FOR I=1 TO VAL(A$(1,9))
1160 PRINT#1,D(1,I)
1170 NEXT I
1180 IF VAL(A$(1,1))=1 THEN 1250
1190 FOR I=1 TO 14
1200 PRINT#1,A$(2,I)
1210 NEXT I
1220 FOR I=1 TO VAL(A$(2,9))
1230 PRINT#1,D(2,I)
1240 NEXT I
1250 CLOSE#1
1260 RETURN
```

Change Directory Path Subroutine

```
1270 REM
1280 REM
1290 REM CHANGE PATH
1300 REM
1310 REM
1320 CLS:PRINT:PRINT:PRINT:PRINT:PRINT TAB(24);"CURRENT PATH IS ";P$
1330 PRINT TAB(24);:INPUT "ENTER NEW PATH ";P$
1340 RETURN
```

Appendix C: Writing Software for Importing Data into Lotus

Set Initial Values Subroutine

```
1350 REM
1360 REM
1370 REM LABELS AND CONSTANTS
1380 REM
1390 REM
1400 L$(1)="DIMENSION"
1410 L$(2)="FILE NAME"
1420 L$(3)="DESCRIPTION"
1430 L$(4)="OPERATOR"
1440 L$(5)="VARIABLE NAME"
1450 L$(6)="UNITS"
1460 L$(7)="HIGH LIMIT"
1470 L$(8)="LOW LIMIT"
1480 L$(9)="COUNT"
1490 L$(10)="TITLE 1"
1500 L$(11)="TITLE 2"
1510 L$(12)="X-TITLE"
1520 L$(13)="Y-TITLE"
1530 L$(14)="FLOATING LABEL"
1540 P$="B:\"
1550 RETURN
```

File Translation Macro

The following Macro listing has been transferred directly from a working spreadsheet and can be entered exactly as it is shown. Required setup commands have been provided earlier in this appendix.

```
J3:       ' ********
K3:-O4:      '*********
J18:-O19:    '*********
P3:-P19:     '**
P4:-P18:     '***
J5:-J17:     '***
K7:       'STATISTICAL MODELING WITH LOTUS 1-2-3
L8:       'COMPANION DISKETTE
K11:      'A STATISTICAL SOFTWARE PACKAGE
K15:      'FILE TRANSLATION TEMPLATE
I21:      'MACRO
I23:      '{home}{goto}i1~/xlEnter name of file to import~i25~/ci25~i27~
I24:      '{home}/ci25~i45~/fit
I26:      '~{right}{right}/fin
I28:      '~/c~{left}{left}~{goto}c7~
I29:      '/c{down}{down}~{left}{left}~
I30:      '{goto}c15~/c{end}{down}~{left}{left}~
I31:      '{end}{down}~{down}{down}{down}
I32:      '{down}{down}{down}{down}
I33:      '/c{down}{down}~{left}{left}~
I34:      '{down}{down}{down}{down}
I35:      '{down}{down}{down}{down}
I36:      '/c{end}{down}~{left}{left}~
I37:      '{goto}a15~{end}{down}~
I38:      '{down}/m.{down}{down}
I39:      '{down}{down}{down}{down}
I40:      '{down}{down}{down}{down}
I41:      '{down}{down}{down}{down}
I42:      '{end}{down}~b1~{home}
I43:      '/rec1.d2047~
I44:      '/fxf
I46:      '~a1.b2047~r~
I47:      '/frmenu~
```

APPENDIX D: MACRO MENUS AND MARQUIS

There are two sections in this appendix. The first is a lighthearted look at writing macros with menus, subroutines, loops, and interactive keyboard input. The ancient game of NIM will be used as a medium, and a program to play the game in Lotus will be developed. It has been included on the companion disk under the name NIM.

The second section of this appendix will take the techniques developed and apply them to an interactive menu package for the templates developed in this book. Those who have purchased the companion disk will find these menus included in each template and may load the statistics package by retrieving the MENU spreadsheet. Whether this menu set is entered by the reader or purchased on the companion disk, the end result is a basic statistics software package that is flexible and fully menu-driven. Those working through the text are encouraged to take the extra step and generate the master menu.

The Game of NIM

NIM predates both chess and checkers, making it one of the oldest games known to man. It is not played as often as these other games because of an inherent flaw; if both players know how to play, the one who goes first always wins. The moves are simple; twenty-one objects are placed in front of the players who, in turn, remove either one or two from the pile. The one who removes the last one loses.

Despite its flaw, this game has been used frequently as a training aid in both hardware and software courses. The purpose here will be to illustrate spreadsheet macro menus (/XM), macro subroutines (/XC and /XR), and inputs to running macros (/XL and /XN).

Figure D.1 shows the screen layout after the NIM spreadsheet has been loaded. Note that the menu active on the command line is not one of the standard worksheet menus but is similar in appearance to those available following a slash (/) input. The similarity continues into operation of the menu. The arrow keys may be used to direct the cursor to the proper selection, or the first letter of the selection may be typed in, as with standard menus.

The following sections describe how to generate this screen, the two menus, the included subroutine, and all the necessary range names.

```
              THE ANCIENT GAME OF NIM

A demonstration of MACRO menus in LOTUS 1-2-3

     Rules:     1. The game begins with a box containing 21 stones
                2. The player decides who moves first
                3. In turn the player or computer remove 1 or 2 stones
                4. He who removes the last stone loses
                5. <ESC> TO HALT, <ALT> A TO REPLAY

Number of stones in the box:              21

                        has removed               stone(s)
            LOTUS       has removed               stone(s)
```

Figure D.1 Game Board Layout

The Screen Layout

The following listing provides the cell locations and contents required to generate the layout in Figure D.1. Remember, if the first character is an apostrophe ('), the contents are treated as a label. Move to the indicated cell and enter the contents exactly as shown; entries spread over two lines must be entered as a single line in the cell even though they run off the screen. Many of the cells overlap, so the completed spreadsheet will be illegible in some areas. Increased use of subroutines and /XG commands increases readability but slows down loading and running, so it has been avoided.

Appendix D: Macro Menus and Marquis

NIM Screen Layout

```
B5:  'A demonstration of MACRO menus in LOTUS 1-2-3
B7:  'Rules:
B13: 'Number of stones in the box:
C3:  'THE ANCIENT GAME OF NIM
C7:  '1. The game begins with a box containing 21 stones
C8:  '2. The player decides who moves first
C9:  '3. In turn the player or computer remove 1 or 2 stones
C10: '4. He who removes the last stone loses
C11: '5.   TO HALT,< ALT/ A> TO REPLAY
C15: '
C16: 'LOTUS
D15: ' has removed
D16: ' has removed
F13: 21
G15: 'stone(s)
G16: 'stone(s)
```

Macro Program Body

The next program segment is the main body extending from B21 to B45. B21 is an inert statement used as a column heading. Cell B22 is actually the first cell in the startup macro. Prior to entering this macro, move the cursor to cell B22 and use the following commands to name it both as range \0 and \A:

/RNC\0<return><return>   (range, name, create, \0)
/RNC\A   (range, name, create, \A)

Appendix D: Macro Menus and Marquis

When the spreadsheet is loaded, the \0 macro automatically executes. Giving the cell a second name (\A) allows the program to be restarted if it is halted.

Cell B22 requests the player's name and stores it in cell C15, using the /XL command. After it is stored, the /RLR (range, label, right) command moves the name to the right side of the cell. Cell B23 is a subroutine call (/XC) that accomplishes the following:

- Cell B24 is identified as the cell following the subroutine call; 1-2-3 stores this location for later use.

- Execution is transferred to the first cell in the range, named Menu (this range is defined in the next section).

- Execution continues from this new location until a /xr subroutine return is encountered.

- When a /xr is found, macro execution returns to the cell just defined, B24, where it continues as if it had never left.

The purpose of the Menu subroutine is to generate a command line that allows the player to choose between removing zero, one, or two stones (zero is only allowed on the first move). The player's choice is recorded at cell F15, and the remaining number of stones is stored in G21.

Cell B24 is the first step executed after the subroutine transfers to the menu. It, as well as the next 20 statements, is an if-then statement (/xi). After the menu subroutine has been executed, cell G21 contains the number of stones remaining. Cell B24 transfers control to F35 if there are less than two stones left. The actual form of the statement is

```
/XIcondition~location~
```

where condition is a mathematical relationship. In this case, the relationship is

```
g21<2
```

(The contents of cell G21 are less than two.) When this condition is met, the end of the program has been reached; the player has won.

The next 20 cells are identical in form and function. Cell B25 may be entered and copied over the range B26-B44, and each cell may be edited to the correct values instead of typing each cell from scratch. When control returns from the subroutine, there will be a value between 1 and 21 in cell G21. Because this value will be a constant, only one of the 21 If statements will actually execute. If that one If statement is B24, the program will end. If it is one of the remaining 20, three possible results can occur:

- The cursor moves to F16 (the display of the computer's last pick), and the value is displayed.

- The cursor is moved to F13 where a value is stored. This value is the sum of the computer's and player's picks so far.

- The cursor is sent to A1.

Appendix D: Macro Menus and Marquis

B45 branches back to the Menu subroutine call for a new pick by the player. This loop is repeated until the player wins (cell G21 has a value of 1 when the subroutine return is executed) or the computer wins. In this case, execution is halted in the subroutine.

```
B21:  'MACRO
B22:  "{home}/xlWhat is your name challenger    ~c15~/rlr~
B23:  '/XCMENU~
B24:  '/xig21~/xgf35~
B25:  '/xig21=2~{goto}f16~1~{goto}f13~1~{home}
B26:  '/xig21=3~{goto}f16~2~{goto}f13~1~{home}
B27:  '/xig21=4~{goto}f16~1~{goto}f13~3~{home}
B28:  '/xig21=5~{goto}f16~1~{goto}f13~4~{home}
B29:  '/xig21=6~{goto}f16~2~{goto}f13~4~{home}
B30:  '/xig21=7~{goto}f16~1~{goto}f13~6~{home}
B31:  '/xig21=8~{goto}f16~1~{goto}f13~7~{home}
B32:  '/xig21=9~{goto}f16~2~{goto}f13~7~{home}
B33:  '/xig21=10~{goto}f16~2~{goto}f13~8~{home}
B34:  '/xig21=11~{goto}f16~1~{goto}f13~10~{home}
B35:  '/xig21=12~{goto}f16~2~{goto}f13~10~{home}
B36:  '/xig21=13~{goto}f16~1~{goto}f13~12~{home}
B37:  '/xig21=14~{goto}f16~1~{goto}f13~13~{home}
B38:  '/xig21=15~{goto}f16~2~{goto}f13~13~{home}
B39:  '/xig21=16~{goto}f16~2~{goto}f13~14~{home}
B40:  '/xig21=17~{goto}f16~1~{goto}f13~16~{home}
B41:  '/xig21=18~{goto}f16~2~{goto}f13~16~{home}
B42:  '/xig21=19~{goto}f16~1~{goto}f13~18~{home}
B43:  '/xig21=20~{goto}f16~1~{goto}f13~19~{home}
B44:  '/xig21=21~{goto}f16~2~{goto}f13~19~{home}
B45:  '/XGB23~
```

Appendix D: Macro Menus and Marquis

The Subroutines

Menu Call

Cells F21 and G21 are not really part of this subroutine. F21 is a label for G21, which will contain the number of stones in the box (less the players pick). F22 is a label for the single-line subroutine — F23 — which calls Menu1 using a Menu Call command (/XM). F23 must be named so when the main body executes the subroutine call, 1-2-3 knows where to transfer control. Before entering these statements, go to F23. Then, enter

/RNCmenu<return><return>

Menu Call Subroutine

```
F21: 'New Count
G21: (F13-F15)
F22: 'Menu Subroutine
F23: '/xmmenu~
```

Menu

F26 is a label for the menu range defined in cells F27 to H28. The labels in F27, G27, and H27 are what is actually displayed on the top line of the command row when this menu is executed. They are the allowable responses executed either by moving the cursor or typing the first letter, as with 1-2-3's normal command menus. The labels in F28, G28, and H28 are the second-line labels displayed when the cursor is highlighting the corresponding first-line label. For the menu to execute, a range must be named before it is used. Go to cell F27 and execute the following:

/RNCmenu1<return>   (range, name, create)
<right><right><down><return>   (F27..H28 highlighted)

When the menu is executed, the player will be offered three choices: zero, one, or two. When the response is given, the macro will continue at F29, G29, or H29. Those three listings follow the actual menu.

Appendix D: Macro Menus and Marquis

Number of Moves Menu

```
F26:  'menu1
F27:  'ZERO
F28:  'Computer moves first
G27:  'ONE
G28:  'Remove one stone
H27:  'TWO
H28:  'Remove two stones
```

Count Submacros

Each of the three following subsections is an extension of one of the menu columns. In the first subsection, F29 declares the player a loser for trying to take zero for other than the first move. Control is transferred to the replay menu. F31 picks two stones for the computer if it moves first.

```
F29:  '/XIf13>1~{GOTO}b18~THAT IS CHEATING -- YOU LOSE!!~/XGF38~
F30:  '{GOTO}b18~/RE~
F31:  '{GOTO}F16~2~
F32:  '/XR
```

The remaining two sections are identical — with one exception. Lines G29-32 process a selection of one stone, and H29-32 processes a selection of two stones.

Value Selection Subroutines

```
G29:  '{GOTO}B18~/RE~
G30:  '{GOTO}F15~
G31:  '1~
G32:  '/XR


H29:  '{GOTO}B18~/RE~
H30:  '{GOTO}F15~
H31:  '2~
H32:  '/XR
```

Each of the three routines ends with a Return From Subroutine command (/XR) because after every selection, control must pass back to the main program body.

Replay Menu

The final listings provide the final routines to complete the package. There are three possible values for the number of stones remaining when control is transferred to F35. If the content of G21 is 1, the player has won; if it is zero, the computer (programmer) has won. If it is -1, the player tried to remove two stones when only one was left.

The If statement in F36 is executed if the player wins; F37 is executed if the programmer wins, and F38 is executed if the player tries to draw two when one is left. As in the main program body, only one of these statements will execute because G21 is a constant.

F38 calls the second menu in this program — the Replay menu.

```
F34:    'ENDIT
F35:    '/xig21=1~{goto}b18~YOU HAVE BEATEN ME
        - YOU MUST HAVE GONE FIRST
F36:    '/XIG21=0~{goto}b18~           LOOKS LIKE I GOT YOU~
F37:    '/XIG21=-1~{goto}b18~          THERE IS ONLY ONE STONE
        - NICE TRY, YOU STILL LOSE!
F38:    '{GOTO}F13~0~/XMREPLAY~
```

F41-H43 is a menu range called from cell F38. The three defined options are as follows: Replay, executed by retrieving the file again to reset parameters; Quit, which executes a macro quit (/XQ); and Exit, which executes /QY to leave Lotus. Don't forget to name the menu range. Move the cursor to cell F41 and enter

/rncreplay<return><right><right><down><return>

Appendix D: Macro Menus and Marquis

Replay Menu and End Subroutine

```
F40:  'REPLAY
F41:  'Replay
F42:  'Play Again
F43:  '/frnim~


G41:  'Quit
G42:  'Quit the game
G43:  '/xq


H41:  'Exit
H42:  'Leave 1-2-3
H43:  '/qy
```

Running the Macro

Two names were given to the beginning range of this macro — \0 and \A. When the NIM file is retrieved, the macro will execute immediately, asking for the player's name and displaying it on the screen. The first menu then offers the choices zero, one, and two on the command line. As stones are removed, the player can watch the command line and follow the macro action as calculations are made and the screen updated. Using step mode (<Alt/F1>), you can reduce this action to a single step.

There are a few screen comments buried in the macro, but they are self-explanatory. When the last play is made, the second menu appears. You should play through several times to get the feel of these menus. Remember, execution can be made through the Arrow and Return keys or by pressing the first letter of the menu command.

Macro execution can be stopped at any time, and <Alt/A> will restart it. Since this macro was kept simple, a restart does not reinitialize variables (it easily could have), so play will be altered if a game is continued after a halt.

Appendix D: Macro Menus and Marquis

The Statistics Package Menu

Whether you buy the companion disk or enter the templates from this book, you will wind up with six spreadsheets encompassing 25 of the most commonly used statistical functions. Chapter 13 summarizes these functions and how they can be used together. This section describes the MENU spreadsheet — an interactive template that allows the user to load the statistics package developed here as a unit and move between functions through the use of menus.

There are several advantages to this interactive capability: it makes the package simpler to learn and easier to teach to new users and encourages experimentation with new methods. The menu has a built-in marquis, as do all the templates described in this text. For new users who are uncertain of the methods used, this is a reassuring entry point to the programs. Figure D.2 shows the structure of this menu template in tree form.

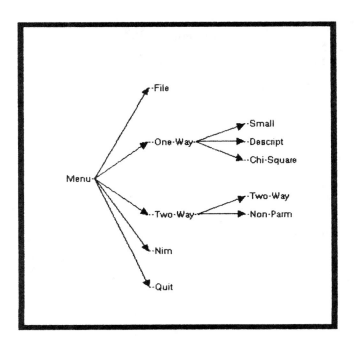

Figure D.2 Statpak Menu Tree

461

Appendix D: Macro Menus and Marquis

The Marquis

Numerous terms are used to describe the introductory screen in a software package; a common one is **marquis**. The first listing that follows is a frame with a small moving graphic to be used in introducing the menu. It can be stored in a separate file and modified for use in the template programs if you want. Remember, if you are developing spreadsheets for other people to use, introductory items such as the marquis are the touches that make users comfortable with your software

The following listing is a template that draws the introductory marquis on the screen.

```
B3,B19: ' *
B4-B18: '**
C3-I4:  '*********
C18-I19: '*********
C5-C17: '*
D11:  '    A STATISTICAL SOFTWARE PACKAGE
D15:  ' FROM: Management Information Systems
D7:   'STATISTICAL MODELING WITH LOTUS 1-2-3
E13:  '         MENU
E8:   'COMPANION DISKETTE
H18:  '*
H19:  '*
H3:   '*
H4:   '*
I5-I17: '           *
J3,J19: '*
J4-J18: '**
```

The listing starting with cell A21 is the macro that automatically executes when the spreadsheet is loaded. The cursor is sent to home and then moved down and across to cell D15. The display is modified by copying the new text entry into that cell. The final line executes the first menu.

Appendix D: Macro Menus and Marquis

The Introductory Marquis

```
A21:  'menu
A22:  '{home}
A23:  '{down}{down}{down}
A24:  '{down}{down}{down}
A25:  '{down}{down}{down}
A26:  '{down}{down}{down}
A27:  '{down}{down}
A28:  '{right}{right}{right}
A29:  '       BY: William T. Cloake III~{home}
A30:  '/xmmenu1~
```

Menus

The the following listing is the first menu active on the screen when the spreadsheet is loaded. Six choices are offered, as was described in Figure D.2. Based on the user selection, macro flow is redirected.

Three of the choices are directly accessible. Selecting **File** loads the data input and update template for disk file maintenance. **Nim** loads the game described earlier in this appendix. **Transl** executes the data translation utility. **Quit** terminates the macro and executes a /qy (quit 1-2-3).

The final two options branch to menus called **One-way** and **Two-way**, and from these menus, the final five templates are loaded. The SMALL template is small sample statistical techniques. DESCRIPT is descriptive statistical techniques. CHI__SQ is similar to DESCRIPT, with some additions. TWO-WAY allows the user to choose between normal two-way techniques and non-parametric methods.

Appendix D: Macro Menus and Marquis

```
D33:  'menu1
D34:  'File
D35:  'Disk File Operations
D36:  '/frdatain~
E34:  'One-Way
E35:  'Small   Descript   Chi-Sq   Quit
E36:  '/xmmenu2~
F34:  'Two-Way
F35:  'Two-Way  Non-Parm  Quit
F36:  '/xmmenu3~
G34:  'Nim
G35:  'Play the game of NIM
G36:  '/frnon_parm
H34:  'Transl
H35:  'Translate File to Lotus Format
H36:  '/frtransl
I34:  'Quit
I35:  'Exit Statistical Analysis
I36:  '{goto}d15~/re~
I37:  '{right}    THANK YOU~
I38:  '{home}
I39:  '/qy
```

The second menu allows the user to retrieve either the SMALL, DESCRIPT, or CHI__SQ spreadsheet or return to the primary menu.

```
D43:  'menu2
D44:  'Small
D45:  'Small sample statistical methods
D46:  '/frsmall~
E44:  'Descript
E45:  'Descriptive statistical methods
E46:  '/frdescript~
F44:  'Chi-Sq
F45:  'Chi-Squared methods
F46:  '/frchi_sq~
G44:  'Quit
G45:  'Return to main menu
G46:  '/xmmenu1~
```

The final menu allows the user to choose between Two-Way and Non-parametric methods or return to the main menu.

```
D49:    'menu3
D50:    'Two-Way
D51:    'Two-way techniques
D52:    '/frtwo_way~
E50:    'Non-Parm
E51:    'Non-parametric methods
E52:    '/frnon_parm~
F50:    'Quit
F51:    'Return to main Menu
F52:    '/xmmenu1~
```

The final menu allows the user to choose between Two-Way and Non-parametric methods or return to the main menu.

```
D49:    'menu3
D50:    'Two-Way
D51:    'Two-way techniques
D52:    '/frtwo_way~
E50:    'Non-Parm
E51:    'Non-parametric methods
E52:    '/frnon_parm~
F50:    'Quit
F51:    'Return to main Menu
F52:    '/xmmenu1~
```

Range Names

Once the complete spreadsheet has been loaded, all that remains is to name the required ranges. The following five command lines will generate the names for the menu and macro start ranges:

```
/rncmenu1<return>d34.h35<return>
/rncmenu2<return>d44.g45<return>
/rncmenu3<return>d50.f51<return>
/rnc\0<return>a22<return>
/rnc\A<return>a22<return>
```

APPENDIX E: CUSTOMIZING PRINTGRAPH OUTPUTS

Chapter 5 provides information on creating and saving graphs. The ultimate result of these operations is a .PIC file that can be displayed. There are many ways to print this file, with the most convenient being Lotus Printgraph. Printgraph is a separate program furnished when you purchase 1-2-3, and, just like the main program, it is loaded through the Lotus access system menu. There are other third-party packages for displaying graphs; some are even three-dimensional.

Figure E.1 shows the command structure for Lotus Printgraph. This appendix will briefly explain the purpose of each of these branches, generate an example graph, and generate a custom output using Printgraph.

Printgraph Functions

Lotus Printgraph releases 1 and 2 perform the same function with quite different command trees. The biggest difference between the two trees is that with release 2, the names of the commands are more descriptive. **Select** is replaced by **Image-Select**; **Options** and **Configure** are replaced by **Image** and **Hardware**; and both are encompassed by the new **Settings** function. Figure E.1 shows these two trees for comparison.

Appendix E: Customizing Printgraph Outputs

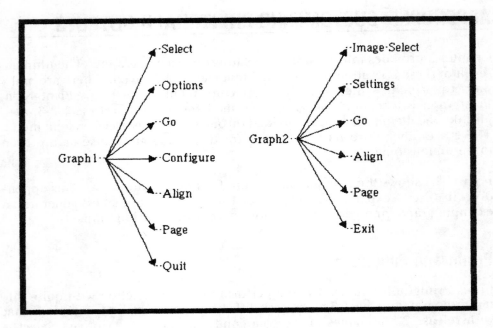

Figure E.1 Printgraph Menu Tree

Despite these differences, the functions of releases 1 and 2 are the same. In fact, graphs created in one release can actually be printed by the other's Printgraph utility. This appendix will concentrate on the release 2 version.

Image-Select allows the user to pick one or more graphs from the active directory to be displayed. The Settings option is divided into two sections: Image and Hardware.

Image allows modification of the display parameters. The size of the graph can be set to either **full**, **half**, or **manual**. The full-page graph is aligned with the paper and turned 90 degrees since the standard graph is wider than it is tall. Half-page graphs are aligned with no rotation and can be printed two to a page. Manual allows custom printing and is the subject of the last section of this appendix, where size and rotation will be explained in depth. There are two font selections allowing the user to specify which of the available type styles will be used to generate the graph (block, bold, forum, italic, Lotus, Roman, and script). **Range colors** allow the user to specify what color the outputs will be on color printers and plotters.

The **Hardware** command allows specification of the mode operation of the program. Graph and font directories allow system configuration. Usually, these directories will be set to A:\ when the system is first installed. For a two-floppy system, the graphs directory should be changed to B:\. For a hard disk system, the fonts directory will probably be the subdirectory in which Lotus 1-2-3 is located. Remember, if you want to reset any directory in 1-2-3 or Printgraph, you must have a disk in the drive (e.g., to change the fonts directory from A:\ to C:\LOTUS\, you must have a disk in drive A:, and the subdirectory must exist on drive C:).

Interface allows selection of one of the available printer or COM ports, which is where graphs will be outputted by this program. **Printer** provides the list of output devices supported and loads the correct driver. The driver is a small program, external to Printgraph, which converts its output into a format that can be recognized by the printer or plotter you are using. Lotus has done a credible job of providing drivers for a wide variety of devices. Finally, **size-paper** allows specification of the type of paper used. Printgraph scales its output accordingly.

Appendix E: Customizing Printgraph Outputs

Sample Graph

The last section of this appendix is a description of how to customize the size and position of a graph on a page using the manual size commands. To illustrate these functions, a sample graph is required. The equation below will be graphed using 1-2-3:

$$Y = \sqrt[3]{x^2} \pm \sqrt{36 - x^2}$$

Equation E.1

Since there is a plus/minus term, there are actually two equations to be graphed. The constant in the second term is only real valued over the range -6 to +6, so this is the range used to generate the graph. The following commands will provide the data table:

```
/dfa1.a25<return>-6<return>
.5<return>6<return>
<F5>b1<return>
(A1^2)^(1/3)
<right>
(($A$25^2)-(A1^2))^0.5<return>
<right>
(B1+C1)<return>
<right>
(B1-C1<return>
/cB1.E1<return>B25<return>
```

Fill a column with -6 to 6.
GoTo B1.
First term in equation E.1.
GoTo C1.
Second term in equation E.1.
GoTo D1.
Equation E.1 first solution.
GoTo E1.
Equation E.1 second solution.
Expand to table.

Having created the table, all that remains is to enter the graph parameters. The following commands will generate the graph and save it under the name "cardioid":

```
/gtxxa1.a25<return>                         Begin graph, set type
                                            X-Yand x range.
ad1.d25<return>be1.e25<return>              Set a and b ranges.
otfGraph of a cardioid<return>              Enter first title line.
tsAppendix E<return>                        Enter second title line.
txX-Axis<return>                            Enter X-axis title line.
tyY-Axis<return>                            Enter Y axis title line.
laTop<return>lbBottom<return>               Enter legends.
qscardioid<return>v                         Save the graph PIC, view.
```

Figure E.2 shows the output as it appears on your screen. This is the sample graph that will be used in the next section. It will be reproduced five different ways using Lotus Printgraph. A distinctive graph was chosen so that you can see at a glance how the different sizes and orientations affect the printed graph.

Appendix E: Customizing Printgraph Outputs

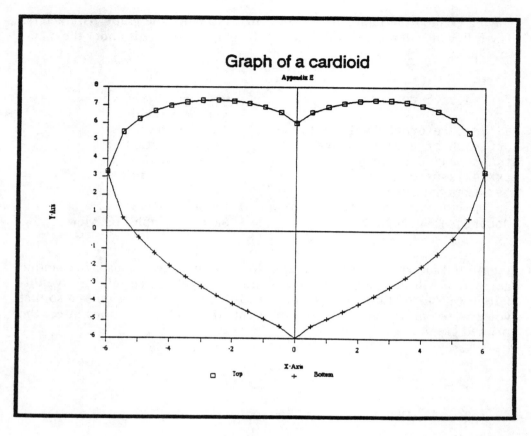

Figure E.2 Test Graph

Manual Print Options with Printgraph

A combination of word processing knowledge and the ability to employ the manual print options (explained next) can yield some striking hybrid reports. The technique to be used is **multiple-pass printing** of the page. On the first printing, text is blocked on the page, leaving space where a graph is to be inserted. The page is then reinserted into the printer, and the graph is added using Printgraph.

Appendix E: Customizing Printgraph Outputs

Figure E.3 was created in this way. The data is the word processor file containing the text for this appendix, and the graph is the sample just created. The printing on the graph tends to be small and hard to read in this application. For that reason, if the graph needs close scrutiny, you will want to print it in a full- or half-page format. For cases where a curve must be shown and reading the scales is not as important, this method of using a graph as an illustration is quite effective.

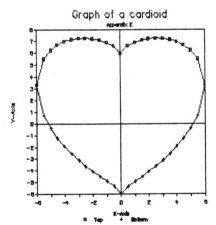

```
Chapter five provided information on creating and saving  graph.
The ultimate result of these operations is a .PIC file which  can
be  displayed. There are many ways to print this file,  with  the
most convenient being LOTUS Printgraph. Printgraph is a  separate
program furnished when you purchase 1-2-3, and is loaded  through
the  LOTUS access system menu, just like 1-2-3. There  are  other
third  party  packages  for displaying graphs,  some  even  three
dimensional.

Figure  E.1 shows  the   command
structure    for      LOTUS
Printgraph. This section  will
briefly explain the purpose of
each  of  these  branches,
generate an example graph, and
generate a custom output using
printgraph.


PRINTGRAPH FUNCTIONS


LOTUS Printgraph releases 1 and 2 perform the same function  with
command trees that are quite different. The biggest difference is
that  with  release  2  the  names  of  the  commands  are   more
descriptive.  Select  is replaced by  Image-Select,  Options  and
```

Figure E.3 Example of Mixed Output

Appendix E: Customizing Printgraph Outputs

The final example has two purposes: it serves as an exercise for readers to test their ability to perform manual manipulations and visually illustrates five different graph formats (there are an endless variety). The following procedure will generate the graph output shown in Figure E.4. It does require that the same sheet be fed into the computer five times. Check closely to see that the paper is aligned to the top of form in your printer, and, if possible, establish a mark so that when you back it up to the top again, you hit the same spot.

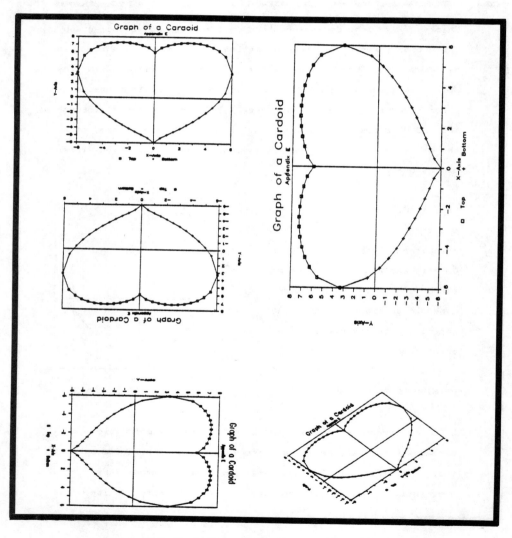

Figure E.4 Variations on Output using Printgraph

Table E.1 has the manual settings for the five graphs. Plot #2 is in the upper left corner of Figure E.4. The next two are below #1. The final 2 (#4 and #5) are in the right-hand column. The best way to understand this section is to work through the instructions that follow, create a copy of Figure E.4 on your own printer, and then review this table again. You will then understand it more fully. This procedure will also give you the opportunity to measure the full-size graph with a ruler and establish whether or not your paper is aligned correctly.

	1	2	3	4	5	Plot
Parameter						
Left	0.25	0.25	0.25	4.25	4.25	
Top	1.00	4.00	7.50	1.00	7.50	
Width	3.50	3.50	3.50	3.50	3.50	
Height	2.50	2.50	2.50	5.50	2.50	
Rotation	0.00	180	270	90	45	

Table E.1
Graph Settings for Final Example

The procedure for the example generation is as follows:

1. Set up your printer and be sure that the paper is at the top-of-form. Load Lotus Printgraph. Make sure the disk with the Cardioid file is in the drive.

2. In Printgraph, select the graph cardioid (<I>mage-select).

3. Enter the manual size settings in column 1 to print the first of the five graphs. The procedure is as follows:

sism10.25<return>	Settings Image Size Manual Left 0.25
t1.00<return>	Top 1.00
w3.50<return>	Width 3.50
h2.50<return>	Height 2.50
r0.00<return>	Rotation zero degrees
qqqqag	Quit Quit Quit Quit Align Go

Appendix E: Customizing Printgraph Outputs

These settings will print the first graph in the upper left corner. In the last command line, the Align is very important. After the printer readjusts the paper, it will give you an automatic form feed after each 11" of paper has been printed; otherwise, the printer would eject the graph before you are done.

4. Repeat Step 3 for the other four graphs, using data from columns 2-5. It does not matter what order you choose.

There are some final observations to be made about the graphs. Obviously, the size is small, but the print remains fairly legible. Graph 2 is upside-down, and 3 and 4 are rotated in opposite directions. If the paper is folded in half and turned 90 degrees (booklet form), these become proper orientations. The proportions of graph 3 have changed significantly. Height and width reverse when the graph is rotated; the height entered is the height of the output — *not* the height of the actual graph. This particular graph looks much better in this format. Number 5 is a nice illustration for a report or a good transparency for overhead projection.

The last exercise is to take your output and a ruler and go back and tackle Table E.1. Graph 1 starts 0.5" from the left margin. The Left entry in Column 1 was 0.25", but the overall display width is actually 8". There is a built-in 0.25" margin on each side. Graph 1 is 3.25" wide, so there is a built-in margin. Just as the width had a built-in margin, the height of the printed area is only 10.5" out of the available 11". The print height is exactly as entered. You can verify the other measurements in a similar manner for your printer.

To actually put the image onto a page, as was suggested previously, you must first print the page with space allowed for the graph. The space is measured and size options are figured from that calculation. The width and height of the space are the same as those required for the graph, allowing a slight margin for error. The top and left are then measured and entered. Rotation usually will be zero degrees in this option. Remember, the Printgraph program will automatically insert 0.25" at the top and left.

APPENDIX F: CREATING MACRO LISTINGS

In any system, especially as complexity increases, the need for quality documentation cannot be understated. This is especially true in software generation. Anyone who has written an application program knows that six months after it is generated, it becomes nearly impossible to remember what each individual line does. The same holds true for macros. It is necessary to keep listings and, where possible, annotate them. Sooner or later, someone will need to fix something or make a change.

The method for listing macros presented in this chapter was used to generate the listings in this book. It requires several steps but allows the programmer to sort the macro listing, annotate it, and print it without retyping. The method is suitable for creating ASCII files with the macro listing. These files can be imported into almost any word processor.

File Printing

Figure F.1 shows a simple spreadsheet that will be used to demonstrate these techniques. Enter it exactly as shown. The result of executing the following commands will be a data file containing a listing of the macro:

```
/pfF_1<return>           Print File
oou                      Options Other Unformatted
ocq                      Other Cell-formulas Quit
r<home>.<end><home>      Range Home..End
gq<return>               Go Quit
```

Appendix F: Creating Macro Listings

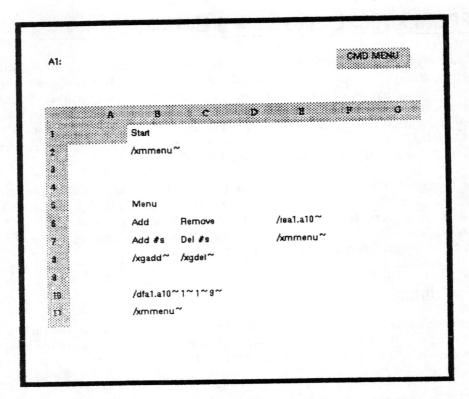

Figure F.1 Sample Menu for Macro Listing Example

The (S)tart command at B2 calls the menu. From the menu, one of the subroutines can be selected. Each subroutine links to the menu. A column of nine integers is either inserted or deleted.

In the second line, the (O)ptions (O)ther (U)nformatted command suppresses page breaks in the created file. The (O)ther (C)ell-formulas command instructs 1-2-3 to generate the output file with the individual cell formulas displayed rather than the actual data values. The result is a print file named F__1 with a .PRN extension.

Appendix F: Creating Macro Listings

The file created is an ASCII text file. Each line contains the address and contents of a specified cell. The contents displayed will be the actual equation in the cell rather than the result of the calculation that is displayed on the screen.

Exit from Lotus, and enter the following to view the contents of this file. Make sure that you start at the DOS prompt (C:).

type b:F_1.prn<return>

Figure F.2 shows the contents as they appear on the screen.

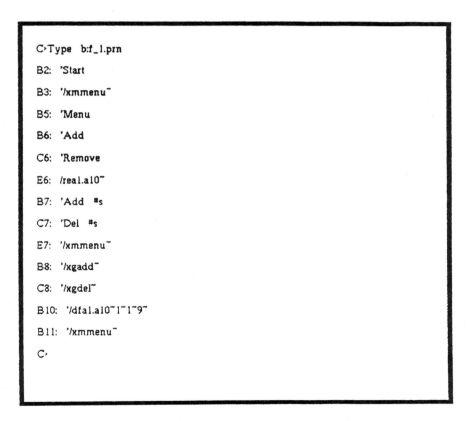

Figure F.2 Macro Listing as Displayed on Screen

Appendix F: Creating Macro Listings

This file can be sent to the printer with the following command:

`copy b:F_1.prn lpt1:`

Importing and Sorting .PRN Files

This file can be imported back into 1-2-3 for examination. Reenter 1-2-3 and execute the following:

`/fitF_1<return>`

Figure F.3 shows the resulting spreadsheet. The ten cells have been reentered, but notice that each has its cell location appended to the contents. The order of cell contents is first numerical and then alphabetical (e.g., A1...An, then B1...Bn). It is now a simple matter to use Lotus to sort these data alphabetically with the following commands:

```
home
/dsd.<end><down><return>           Data Sort Data Range
p.<end><down><return>a<return>     Primary key Ascending
g                                  Go
```

Appendix F: Creating Macro Listings

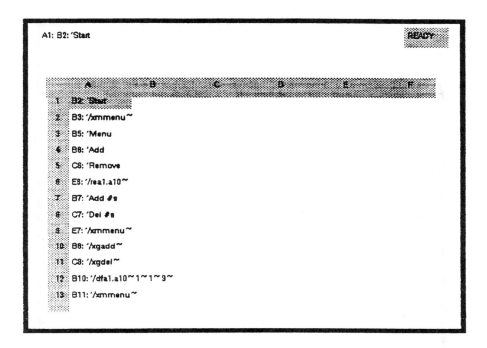

Figure F.3 Sample Template as Printed to File

Figure F.4 shows the result of this sort. The cell contents have been alphabetized, including the numbers. You will see that in a computer sort, B11 comes before B2, so the order may not be as you expected. With a little practice, it will come naturally.

Appendix F: Creating Macro Listings

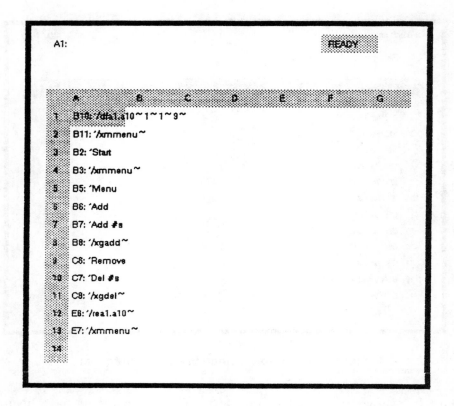

Figure F.4 Data Sorted Alphabetically

Using the Move command, the order shown in Figure F.5 can easily be obtained. In many cases, this will be an acceptable order, but, for the purpose of this exercise, one more modification will be made.

Appendix F: Creating Macro Listings

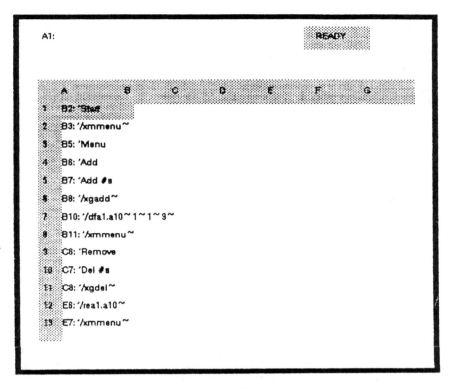

Figure F.5 Data Readjusted after Sort

Appendix F: Creating Macro Listings

Clear the spreadsheet, and reload or reenter the original spreadsheet as shown in Figure F.1. Another data file will be created with range order considered. Enter the following commands:

```
/pff_2oouocq                as above
rb2.b3~g                    print first range
11                          insert two line feeds
rb5.b9~grc6.c8~g            print second range
11                          insert two line feeds
rb10.b11~g                  print the last range
11
re6.e7~9q
```

Erase the worksheet, and import F__2 by entering

```
/fitF_2~
```

Figure F.6 shows the result of these operations. The statements have been stored in the correct order. The created file is an ASCII text file that provides the user with several options. Lotus can be used to print and annotate these archive files, which can be listed and stored like any other spreadsheet. They can also be edited or sorted, and they can be used to rapidly reenter lost templates.

Appendix F: Creating Macro Listings

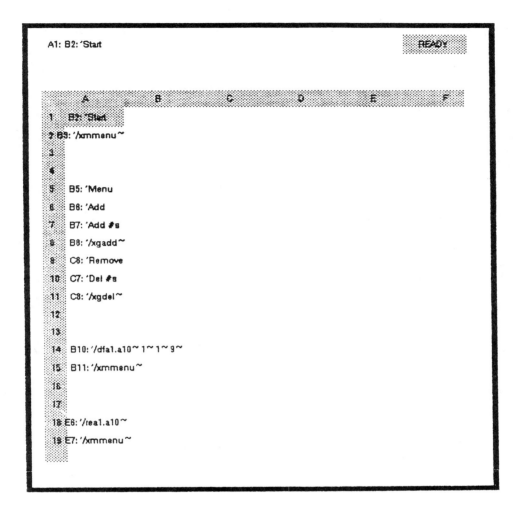

Figure F.6 Output of Macro Template in Sections

The final option is entering the archive files into other applications programs.

As you can see, this method provides not only the capacity for documentation but the ability to generate and use listings in conjunction with other text packages.

APPENDIX G: DOS UTILITIES YOU NEED TO KNOW

It is all too common for users to buy a PC or new version of DOS, back up the system disk, and continue to use the same functions they have always used. Many people only wonder why all 30 or 40 programs cluttering the directory are there. Most applications programs, like Lotus, have an installation utility that allows users to transfer the necessary utilities to the application disk without delving deeper into what those utilities do, which only makes it easier for users to avoid learning some useful tools that can be applied in surprising places.

Most users will never need all these utilities, and very few need to understand them all. There are, however, a few that will make their lives much easier. Many frustrations common to users can be easily solved with tools they have available. This appendix will provide an introduction to some of the more basic DOS tools.

The following three sections each cover a small group of utilities. The first group contains elementary file and disk manipulation utilities: FORMAT, CHKDSK, DELETE, ERASE, COPY, and DISKCOPY. The second group contains advanced directory commands: DIR, MKDIR, CHDIR, RMDIR, TREE, LABEL, and PATH. The final commands are handy for hard disk users: PROMPT, SUBST, and ATTRIB.

File Manipulation

Each of the file manipulation commands has a function related to creating files, moving files, or preparing disks to accept files.

FORMAT

The FORMAT utility is the most basic DOS utility; a diskette is not useable until this utility has been run. The FORMAT utility checks the diskette tracks for defects, creates a directory, and sets up a file allocation table. These tasks actually have a simple function.

Appendix G: DOS Utilities You Need to Know

Standard PCs and compatibles contain 360-kilobyte floppy disk drives. There are two recording heads in each drive that can be positioned to one of forty different radii (called **tracks**). Each track is electronically divided into 9 sections, which means there are 2 * 9 * 40 = 720 places in which data can be stored on the disk. Each place, or **sector**, holds 512 bytes of data, making a total of 512 * 720 = 368,640 bytes on a floppy disk. The system occupies some of this space, but DOS allocates the remaining bytes to the files you store as you create or load files on the disk.

The directory is a disk file itself and includes a table with a list of files appearing on the disk. On a standard diskette, 112 files can be listed in the root directory. The file allocation table is a record, stored on the disk, of what is stored in the 720 locations. The directory and the **file allocation table** (FAT) work together to allow you to save and recover data. Creation of these tools is the primary function of the FORMAT utility. A marker is placed at the beginning of each sector, allowing the system to find that sector and link the stored data via the FAT.

The FORMAT utility is executed by placing a blank disk in one of the drives, typically drive B, and entering

FORMAT B:<return>

or

FORMAT B:/S<return>

In either case, the user is prompted to install a new diskette and press Return. If a diskette already containing data is used, all data will be lost as the tables are recreated. If the /S option is executed, the system files will be transferred as the formatting takes place. In this case, the disk will be bootable (in other words, usable for computer startup). One additional useful option in formatting, /V, allows the user to give the diskette an eleven-character name, which is displayed when a directory command is performed.

CHKDSK

After formatting (or any time you want a status report on the disk), a CHKDSK command can be run. This utility provides a status report on system and disk memory availability. A CHKDSK scans the directory and FAT (file allocation table) and checks internal hardware.

The syntax for this utility is

CHKDSK B:<return>

After a few seconds, the screen will display the status of both disk and RAM. The following display is a typical example:

```
362946 bytes total disk space
194560 bytes in 38 user files
167936 bytes available disk space

655360 bytes total memory
246528 bytes free
```

The 360K diskette in this example is slightly more than half full. The 640K system RAM is more than half used. The 655,360 is called 640K because 1K (kilobyte) in computer terms is 2 raised to tenth power, or 1024 bytes. There are only 362,000 bytes of disk space listed instead of 368,000 as calculated above. The missing 6K holds the directory and FAT.

DISKCOPY

DISKCOPY allows complete duplication of a floppy disk. The syntax for this utility is

DISKCOPY A: B:<return>

The entire contents of diskette A are transferred to diskette B. DISKCOPY is a mirror copy utility that will format the diskette if necessary. It will not work for copying to or from a hard disk, but the DISKCOPY program can be stored on a hard disk.

COPY

The COPY utility is more adaptable than the DISKCOPY utility. Some or all of the files can be moved, and transfers can be made to and from hard disks. Also, the new disk need not be an exact mirror image of the old.

The advantage of this nonmirror copy is subtle but powerful for applications program users who have a large number of data files. As spreadsheets are loaded and stored on a data disk, they are placed on the disk sequentially. The drive heads move to the data file and step sequentially through the disk when the spreadsheet is being loaded.

If one or more small spreadsheets are deleted, however, nonsequential gaps with the potential to disrupt the efficiency of this nonmirror copying capability can be created on the disk. Any time a program is stored, it is placed in the first available gap. If a large spreadsheet is stored, it will be placed into these gaps. If it is larger than the first available space, it is linked to the next space as well. Large spreadsheets, then, can wind up spread over nonsequential gaps in a diskette. In loading such spreadsheets, each extent, or jump, requires additional time because the drive heads must be moved and the new sector found. If you have ever experienced a program loading at a speed slower than that to which you are accustomed, you may be having this segmenting problem. Listen to your drive; it will tell you when you have a problem. If you can hear it step (a light clicking sound) at irregular intervals while a program is loading, the program is probably segmented.

Copying the disk with DISKCOPY will not help with segmenting problems because the mirror-image copy will be identical. Using COPY to transfer each file onto a blank disk causes contiguous (nonsegmented) block files to be created. Each file is transferred completely before the next one is copied. This method works on both hard and floppy disks. Hard disk managers that automatically perform this function are available because segmenting can be a severe problem.

Appendix G: DOS Utilities You Need to Know

The basic syntax for the COPY utility is

COPY *from to*

where *from* and *to* are file specifications. If *to* is omitted, the program copies data from the specified location to the current disk and gives it the same name.

Both *from* and *to* in this case have a wide range of allowable syntax as file specifications. Each can contain a path, a file name, and an extension. The asterisk (*) can be used as a wild card for copying multiple files with similar characteristics.

The following examples should help you understand this process:

Command	Function
COPY A:*.* b:	Copy all files from drive A to B.
COPY A:*.BAS b:	Copy all files with .BAS extension from A to B.
COPY B:MYFILE.*	Copy all files whose name is "myfile" with any extension from drive B to the currently active drive.
COPY a:*.PIC C:	Copy all *.PIC files to drive C.
COPY C:\SYSTEM*.*	Copy all files from the system subdirectory of drive C to the current drive.

Table G.1

The first command in Table G.1 will solve the nonsequential gap problem previously described. If the data disk is a floppy, place it in drive A. Place a blank formatted disk in drive B. Type

COPY A:*.* B:<return>

The resulting data disk will have all contiguous data files. This method of copying is a very convenient way to ensure regular backup of files and efficient program loading.

A hard disk management utility is required to ensure regular backup of files and quicker program loading on a hard disk. Even if you empty a subdirectory and reload it one file at a time, the FAT will allocate files to available segments. The new directory may be as segmented as the old one. It is necessary to clear all data off the first few tracks and begin from there to reload the data. A good disk manager will, in the process, put all directories together in the beginning of the disk, which speeds directory searches tremendously.

DELETE and ERASE

Both of these commands remove a file or files from the directory. The wild cards shown previously allow multiple-file operation. The syntax for the DELETE and ERASE commands, respectively, is

DEL MYFILE.BAS

or

ERASE MYFILE.BAS

to erase a file named MYFILE with an extension of .BAS. If wild cards such as *.* are used, the user receives the prompt,

Are you sure (Y/N)?

to avoid accidental erasures.

One final point about these commands: they *do not* erase or delete files. These commands only remove the file name from the directory and make the disk space available in the FAT. There are programs, such as The Norton Utilities, that allow recovery of these "erased" programs, but such programs only work if no additional files have been stored in the data space.

Advanced Directory Utilities

The following sections will explain how to examine the contents of a directory, how to subdivide directories, and how to customize DOS to work with these subdirectories. DIR allows inspection of any directory; MKDIR, CHDIR, and RMDIR allow creation and use of subdirectories; PATH allows attachment of a subdirectory to the current directory; and TREE allows inspection of the directory tree.

DIR

The DIRECTORY or DIR command displays a list of the files on the current diskette or hard disk path. It also allows a file path or wild cards to be requested, as described previously. The syntax

```
DIR C:\SYSTEM\*.COM
```

would display all files on the subdirectory SYSTEM with a .COM extension. Two options make the display of larger directories easier to handle. The /W option,

```
DIR /W
```

causes an abbreviated display of file names in five columns. This option is handy when a large number of files are contained in the directory. The /P option is similar but displays the entire directory listing. When the screen fills, the listing halts until the user presses a key to continue.

Directory Trees

DOS subdirectories can be arranged in a tree structure. This is very similar to the command structure form in 1-2-3. Figure G.1 shows a typical hard disk structure. The main directory branches into several subdirectory paths. The commands in the next section allow you to create and use these subdirectories. The TREE command discussed in this section displays the contents of these paths.

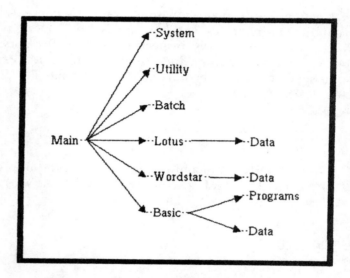

Figure G.1 Sample DOS Menu Structure

TREE

TREE is the only DOS command that seems inadequate for its task. It displays the contents of all of the paths and files, but not in a concise fashion. There are many third-party programs that do a better job. The syntax

```
TREE C:/F
```

displays the contents of each path, starting from the root directory. If you precede this command with a <Ctrl/PrtSc>, the results will be printed as well as displayed on the screen. Repeat the command <Ctrl/PrtSc> to cancel the printer echo.

Subdirectory Operations

This section covers the creation and use of subdirectories. They can be generated by MKDIR, accessed by CHDIR, and removed with RMDIR. The PATH command allows them to be attached to the current directory.

MKDIR

The MKDIR command allows creation of a subdirectory. Surprisingly, this command can even be executed on floppy disks with DOS 2.0 or later. The syntax for creating a subdirectory named "system" on drive C is

```
MD SYSTEM<return>
```

or

```
MKDIR SYSTEM<return>
```

The directory entry for such a subdirectory has no extension. If DIR is executed on the root, the subdirectories each have a <dir> entry attached. This entry does not appear when a DIR is executed with the /W option.

Each subdirectory entry created takes up one file space, so the limit on the number of subdirectories is the limit of files in the current directory. The root on a hard disk can have 512 entries, so it is unlikely that you will use them all.

CHDIR

Once a subdirectory has been created, it must be accessed. The CHANGE DIRECTORY or CHDIR command allows this to happen. To access the system subdirectory just created, the syntax is

CD SYSTEM

or

CHDIR SYSTEM

At this point, if the user typed

DIR<return>

there would be no files except the default routing files:

.

and

..

This subdirectory can be accessed and used just like a floppy or a hard disk root directory. Its path is C:\SYSTEM, just like a floppy in drive B, whose path is B:\. Files can be copied to and from this directory. When the directory is currently operating on the files, it contains no path prefixes.

RMDIR

Before a directory can be removed, it must be empty. Type

DEL *.*

to empty the directory. Then, type:

cd ..
RMDIR SYSTEM

or

cd ..
RD SYSTEM

to remove it.

Appendix G: DOS Utilities You Need to Know

PATH

The last command in this section is highly useful. The PATH command allows attachment of one or more designated subdirectories to the current directory.

The syntax for the PATH command (using the subirectory called "system") is

PATH C:\SYSTEM

With this command active, users can employ FORMAT, DISKCOPY, or any other DOS system command without moving to the subdirectory in which these commands exist.

A second routine application of this command is in running compilers. When working in a subdirectory that contains a compiler or other application program, users often find it helpful to invoke a word processing program. Word processors are useful for generating files that can be fed to applications programs. To override the boot definition system, users can type the path to a word processing program. A path to Wordstar for example, would resemble the following:

PATH WS4<return>

From that point on, if a file must be edited, the user can type WS (using the WordStar example) and invoke the word processor without moving to its subdirectory. Rebooting the system after finishing with that file reinvokes the standard autostart options.

Additional Utility Functions

Three additional functions are covered in this section: PROMPT, SUBST, and ATTRIB. The PROMPT utility allows the user to display the current active DOS path every time the system prompt is displayed; ATTRIB allows setting of file parameters, and SUBST allows a drive specifier to be substituted for a system path.

PROMPT

There are numerous prompts allowable for a hard disk user. The system prompt is normally D, where D is the active drive designator. The command

```
PROMPT $p$g<return>
```

executed either manually or as part of a batch file, will provide the most useful prompt. The $p displays the current path including subdirectory, and the $g is the standard angle bracket (>).

The DOS manual contains a full list of options available for the prompt, including date, time, DOS level, and many others.

Appendix G: DOS Utilities You Need to Know

ATTRIB

The ATTRIB command allows the user to make a file available for reading only. The command:

```
attrib +r myfile.txt
```

makes MYFILE a read only file. If it is accessed by an application program, such as 1-2-3 or a word processor, it can be loaded and viewed only. Changes will not be saved to the file. The command

```
attrib -r myfile.txt
```

cancels this command. The final command

```
attrib myfile.txt
```

displays the attributes of MYFILE.TXT.

SUBST

This is the command for users who have ever wished they could make 1-2-3 support subdirectory paths so they wouldn't constantly having to retype long path names.

Appendix G: DOS Utilities You Need to Know

For data files stored in a subdirectory whose path is

`c:\lotus\data`

the command

`subst e: c:\lotus\data`

will allow you to set the 1-2-3 path to E:\ and access these files. The path name E can be used by DOS, 1-2-3, or any other application program. File actions will be performed on the subdirectory c:\lotus\data. From the DOS prompt, A>, a CD command will transfer control to the subdirectory. DIR will list files just as it will for any other directory.

The command

`SUBST<return>`

will list all active substitutions. These substitutions can be made manually or contained in batch files. If the last command in the batch file (like AUTOEXEC) is SUBST, the screen will show all active substitutions.

To remove a substitution, type

`subst e: /d`

Appendix G: DOS Utilities You Need to Know

DOS allows any letter (up to E) to be used as a substitution by default. In most systems, A-C are designated disk drives, so there are only two available SUBST names. Modification of the CONFIG.SYS command allows characters through Z to be used in this command. To allow any single letter to represent a path, create (or modify an existing) CONFIG.SYS file on the root directory of a boot disk. The command

LASTDRIVE=Z

allows characters through Z to be used in SUBST commands. The CONFIG.SYS file can be generated or modified by most editors (see your DOS manual for details).

INDEX

/file combine, 72
/x commands, 64
/xl, 64
/xn, 64
@sin, 14
@avg, 14
@count, 14
@function, 13
@max, 14
@min, 14
@mod, 14
@pi, 14

A

Additional utility functions, 499
Adjust, 195, 233, 273
Advanced directory utilities, 493
Analysis, 152, 197, 270
Analysis two-way, 266
Anchor, 8
Apostrophe, 61
Attrib, 500
Autostart macro, 128, 133

B

Bar graphs, 95
Bell-shaped curve, 229, 235
Biased Standard Deviation, 398
Bimodal, 157
Bin contents, 202
Bin size, 209

C

CHI_SQ, 399
Carriage controls, 423
Cell boundaries, 190, 266
Cell limit, 169, 202, 229, 236
Cell range names, 140
Cell size, 192, 202
Cellular calculation, 155
Chdir, 496
Chi-Square, 229, 247, 387
Chi-squared distribution, 228
Chkdsk, 489
Coefficient of correlation, 283, 310
Coefficient of determination, 281, 183
Combine, 78, 138, 142
Command line, 128
Composite menu, 386
Contiguous graph points, 93
Copy, 489
Copy entire file, 78
Copy wildcards, 491
Count macro, 136
Criterion, 37
Critical value, 277, 279, 293
Cursor, 195, 233, 273
Curve fit, 280
Curve smoothing, 310
Customizing the package, 389

D

DATAIN, 387
DBMS, 26
DESCRIPT, 228, 387, 397
Data query, 36
Data analysis, 152
Data base, 26
Data base commands, 26
Data distribution, 209
Data entry, 141

Data file, 124, 126, 394
Data fill, 28
Data query extract, 41
Data sets, 26, 124
Data sort, 32
Data table, 44
Datain, 125
dBase, 435
Debug, 56
Delete files, 492
Delimiters, 82
Dependent variables, 266
Descript template, 190
Dimension, 26, 133
Dir /p, 493
Dir /w, 493
Directory Trees, 494
Disk file creation, 394
Disk file options, 391
Diskcopy, 489
Distribution independent, 350
Dos copy, 80

E

EXTRACT, 396
Edit key, 10
Edlin, 80
Equations, Descriptive Statistics, 198-9
Equations, Graphic Methods, 235, 237
Equations, Nonparametric, 354-5, 357-8, 361-2
Equations, Small sample, 167, 169
Equations, Two-way analysis, 277-8, 280
Equations, Two-way graphic, 317-20
Erase, 137
Error margins, 277
Escape sequences, 428
Expected values, 247
Experimental tails, 276
Exploded pie chart, 115
Exponential curves, 310

Extract, 41, 72, 135, 142
Extract values, 76

F

F test, 267, 270, 279, 290, 294, 399
Fast access, 8
Fields, 26, 32
File delete, 492
File print, 77
File combine, 138, 142
File creation, 394
File data, 126
File extensions, 100
File extract, 74, 135, 142
File format, 487
File import, 81, 139, 142
File information, 133
File load operation, 161
File manipulation, 487
File retrieve, 142
Find, 37
Floating labels, 104, 126
Floppy disks, 391
Font sizes, 425
Fonts, 423
Forecasting, 267, 399
Form feeds, 423
Formatting diskettes, 487
Frequency distribution, 157, 174, 190, 211, 228, 245
Frequency distribution table, 203, 229

G

Gauss subroutine, 326
Gaussian, 214, 228
Gaussian curve intersections, 321, 331, 411
Gaussian curves, 282, 318
Gaussian data set, 247
Gaussian Plot, 229

Gaussian plots, 233, 241, 310, 317
Gaussian probability density function, 235
Granularity, 134-5, 190, 229
Graph, 90
Graph, Cartesian coordinates, 266
Graph data labels, 104
Graph, Gaussian, 229
Graph, gaussian
Graph, gaussian intersections, 331, 411
Graph grid, 103
Graph information, 133
Graph labels, 126
Graph line, 243
Graph line labels, 113
Graph min-max-avg, 159, 405
Graph name create, 107
Graph name delete, 107
Graph name use, 107
Graph regression, 314, 327, 330, 409
Graph reset, 100, 107
Graph save, 100, 244
Graph scale, 104
Graph scale skip, 104
Graph scatter, 266, 314, 322, 325
Graph smoothing, 315, 325
Graph title lines, 126
Graph type, 92

H

Halt, 233
Hard disk systems, 392
Headings, 424
Histogram, 190, 202, 228, 247
Hypothesis, 275

I

IMPORT, 72, 78, 139, 142, 396, 435, 480
Increment, 29
Informational, 26
Inputs to macros, 64
Installing the program, 388

K

Kurtosis, 190, 201, 213, 228, 236, 247, 398

L

Large Sample techniques, 397
Lastdrive, 502
Leptokurtic, 201, 214
Line graph, 92, 243
Linear regression, 280
Linear relationship, 266, 283
Listing macros, 477
Low cell, 192
Lprint, 428

M

Macro, 56
Macro cursor, 19995
Macro listings, 477
Macro loop, 68
Macro menu, 134, 193
Macro regression, 287
Macro restart, 137
Macro translation, 449
Main menu, 461
Mann-Whitney Test, 350, 358, 367, 370, 400
Manual graph sizes, 469, 475
Marquis, 128, 451, 462
Mean, 152, 190, 197, 398

Median, 152, 154, 166, 175
Menu adjust, 195
Menu composite, 386
Menu main, 461
Menu menu, 388
Menu printgraph, 468
Menu template, 232, 312, 351, 588
Menus, 451
Mesokurtic, 201, 204
Min-max-avg graph, 159, 405
Mkdir, 495
Modality, 152
Mode, 152, 157, 166, 175
Modified marquis, 389
Modulo, 15
Moment of inertia, 201
Moments, 201
Multi-strike, 423
Multimodal, 157, 159
Multiple fields, 82

N

N-1weighting, 398
NIM, 387, 451
NON__PARM, 387
NUM, 9
Nonparametric methods, 350, 400
Nonrandom patterns, 350
Normal, 214
Normalized counts, 202

O

OGIVE, 228, 233, 239, 247
ONE__WAY, 463
One-tailed, 275

P

PDF, 199, 228, 235
PLOTD, 239
PROJECT, 267, 281, 399
Paint, 8
Parabola, 310, 314
Parse, 80, 82
Path, 498
Peakedness, 228
Pearsonian coefficient of skewness, 169
Pie graphs, 97
Platykurtic, 201, 214
Power curves, 310
Primary key, 33
Print bold, 423
Print enhanced, 423
Print file, 72
Print from macro, 67
Print options, 425
Printer commands, 426
Printer options, 432
Printgraph customizing, 467
Printgraph functions, 467
Printgraph menu, 468
Probability, 197, 317, 328
Probability density function, 199, 228, 235, 310
Prompt, 499
Pseudo data set, 236

R

REGRESSION, 280
Range, 8
Range descriptions, 8
Range names, 11, 140
Range, multiple, 10
Rank sum, 359, 368
Raw data values, 394
Recalculate 69, 195, 206
Record, 26, 32, 82

Regression, 44, 266, 270, 280, 320
Regression Equation, 282
Regression calculations, 310
Regression graph, 327, 330, 409
Regression information, 282, 292
Regressmion macro, 287
Regression plots, 314
Rejection area, 276
Resolution of measurement, 154
Restart macro, 137
Retrieve, 72, 78, 142, 395
Rmdir, 497
Running the package, 394
Runs test, 350, 353, 366, 370, 400

S

SMALL, 152, 387
SMOOTHING, 316
Sample Statistics, 281
Sample size, 152, 199, 202
Scatter plot, 266, 310, 314
Scatter subroutines, 325
Scratchpad, 153, 175, 324
Seasonal variations, 310, 317
Second order relationship, 266, 310, 314
Secondary-key, 34
Setup strings, 431
Sign test, 350, 356, 370, 400
Single-tailed, 275, 294
Skewness, 190, 201, 228, 236, 247, 398
Small sample techniques, 396
Small template, 152
Small template menu, 162
Smooth subroutine, 325
Software for importing, 435
Sort, 233
Sorting macro listings, 480
Stacked bar graph, 92, 96
Standard deviation, 152, 156, 190, 197, 398
Standard error of estimate, 281, 282, 292, 315

Standardized normal random variables, 311
Statistical moments, 190, 201
Step Size, 29
Step function, 56
Stop value, 29
Subdirectory operations, 495
Submacro operation, 135
Submacro, 126
Subst, 500
Sum, 8

T

T Test, 267, 270, 279, 290, 294, 399
TRIM, 396
TWO__WAY, 387, 399, 463
Template, 124, 126
Template Layout, 127, 196, 234, 273, 313, 352
Template Menu, 129, 162, 193, 232, 272, 312, 351, 388
Templates, 16
Test, 273
Test, F, 267, 270, 279, 290, 294, 399
Test, Mann-Whitney, 350, 358, 367, 370, 400
Test, T, 267, 270, 290, 294, 399
Test, Z, 267, 270, 277, 289, 294, 399
Test distribution ind, 350
Test, nonparametric, 350
Test, rank sum, 359
Test, runs, 350, 353, 366, 370, 400
Test, sign, 350, 356, 370, 400
Tilde, 59
Title lines, 126
Translation macro, 449
Tree, 495
Trend indicator, 282
Trim, 137, 142
Two-tailed, 275, 294
Two-way analysis, 266, 270

U

Unbiased standard deviation, 199, 320
Underline, 423
Unique, 44

V

Variable information, 133
Variance, 156, 190, 398
VisiCalc, 435

W

WIPE, 396
Wipe, 137, 142

X

X_Y graph, 92, 98
Xbar and r, 159

Z

Z test, 69, 267, 270, 277, 289, 294, 399

ORDER FORM FOR
PROGRAM LISTINGS ON DISKETTE

*T*his diskette contains the complete program listings for all programs and applications contained in this book. By using this diskette, you will eliminate time spent typing in pages of program code.

If you did not buy this book with diskette, use this form to order now:

Only:

$20.00

MANAGEMENT INFORMATION SOURCE, INC.
P.O. Box 5277 • Portland, OR 97208-5277
(503) 222-2399

NAME (Please print or type)

ADDRESS

CITY   STATE   ZIP

Call free
1-800-MANUALS

☐ *Statistical Modeling disk only $20.00*
Please add $2.00 for shipping and handling.
Please check
☐ VISA   ☐ MasterCharge   ☐ American Express
☐ Check enclosed $ _____

ACCT.

EXP. DATE

SIGNATURE

MIS: PRESS

MANAGEMENT INFORMATION SOURCE, INC.

RELATED TITLES FROM MIS:PRESS

Statistical Modeling Using Lotus 1-2-3
A time-saving guide illustrating how to build statistical models and templates in Lotus 1-2-3 spreadsheets. Complete, step-by-step instructions for entering data from keyboard or file, importing and exporting data from the spreadsheet, passing data from model to model, and producing concise, easy-to-understand output. Sample problems and solutions are provided.

Will Cloake 0-943518-32-6 $22.95 w/disk $42.95

Linear and Dynamic Programming with Lotus 1-2-3
Discusses two of the most useful methods of decision optimization—finding the best alternative among all possible scenarios—within the Lotus 1-2-3 spreadsheet environment. The basic concepts of Linear and Dynamic Programming are explained and illustrated with numerous examples from business and industry. Written for the serious Lotus 1-2-3 user.

James Ho 0-943518-72-5 $19.95 w/disk $39.95

Construction Management Using Lotus 1-2-3
Describes in detail the logic of each phase of the construction management process, including discussions of bidding, estimating job costs, managing cash flow, monitoring equipment ownership and maintenance, and scheduling subcontractor payments. Self-contained, specific applications serve as valuable, easy-to-use guides for anyone involved in the decision-making processes of construction management.

Jay C. Compton 0-943518-17-2 $29.95 w/disk $44.95

Running 1-2-3
An easy-to-use, keystroke-by-keystroke tutorial hailed for its broad repertoire of useful applications. Its popularity with the business and educational communities has made it a proven seller.

Robert Williams 0-943518-78-4 $19.95 w/disk $34.95

Hard Disk Management for IBM PS/2 and Compatibles
Updated for the IBM PS/2 and compatibles, this second edition offers a clear and concise explanation of how to use popular software applications packages on a hard disk. Includes detailed techniques for mastering DOS, installing the programs, writing the menus, managing the memory, finding the files you need, and integrating all these packages into a powerful management tool.

Emily Rosenthal and Ralph Blodgett 0-943518-82-2 $21.95

Advanced DOS: Memory-Resident Utilities, Interrupts, and Disk Management with MS- and PC-DOS
An indispensable resource for serious DOS programmers. Includes chapters on disk data storage, BIOS and DOS interrupts, utility programming, and memory-resident utilities. Addresses topics of DOS and multitasking, subprograms and overlays, writing pop-ups, and reentrancy, and provides step-by-step exploration of the partition table, boot record, traverse paths, and more.

Michael Hyman 0-943518-83-0 $22.95 w/disk $44.95

Running PC-DOS 3.3
An essential desktop reference guide for all PC users. Easier to use and understand than the DOS manual, this book's clear organization, concise instructions, and illustrations help users make the most of their computer's operating system. Includes version 3.3.

Carl Siechert and Chris Wood 0-943518-47-4 $22.95

*A*vailable where fine books are sold.

MANAGEMENT INFORMATION SOURCE, INC.
P.O. Box 5277 • Portland, OR 97208-5277
(503) 222-2399

Call free
1-800-MANUALS

MIS:PRESS

MANAGEMENT INFORMATION SOURCE, INC.